Strange Bedfellows

Strange Bedfellows

Adventures in the Science, History, and Surprising Secrets of STDs

Ina Park

FLATIRON
BOOKS
NEW YORK

STRANGE BEDFELLOWS. Copyright © 2021 by Ina Park. All rights reserved. Printed in the United States of America. For information, address Flatiron Books, 120 Broadway, New York, NY 10271.

www.flatironbooks.com

Designed by Devan Norman

The Library of Congress Cataloging-in-Publication Data is available upon request.

ISBN 978-1-250-20662-6 (hardcover)
ISBN 978-1-250-20665-7 (ebook)

Our books may be purchased in bulk for promotional, educational, or business use. Please contact the Macmillan Corporate and Premium Sales Department at 1-800-221-7945, extension 5442, or by email at MacmillanSpecialMarkets@macmillan.com.

First Edition: 2021

10 9 8 7 6 5 4 3 2 1

To my parents,
James and Young Park

Contents

Contents

Author's Note

The names and some identifying details and characteristics of patients and colleagues who shared their personal stories have been changed to protect their privacy. These persons will be referred to by a first name only, and these names are pseudonyms. Other scientists and colleagues who provided their consent to be identified will be referred to by their real full names.

Although I am a medical doctor with an expertise in many of the diseases under discussion, the general information in this book is *not* intended to replace the particular advice of the reader's own physician or other medical professionals. Readers should consult a medical professional in matters relating to their health, especially if they have existing medical conditions, and before starting, stopping or changing the dose of any medication they are taking, or of any regimen they are following, under the guidance or supervision of their own doctor or other medical professional. Individual readers are solely responsible for their own health care decisions.

Strange Bedfellows

Introduction

Beginning with a Bang

Sometimes life's unhappy accidents inadvertently lead to happy consequences. *Strange Bedfellows* represents the silver lining in a dark cloud that descended on my family in January 2015, when my husband and I experienced a parent's worst nightmare. As we walked out of our house on our way to a birthday party one afternoon, my seven-year-old son, Nate, released my hand and bolted across the street, only to be struck by an oncoming car.

I recall seeing his legs splayed underneath the vehicle and hearing his screams echoing down the street. I, on the other hand, did not utter a sound. This was no longer my son lying on the street—this was a trauma victim who required attention. I ran over to him, silently reviewing the algorithm that was ingrained in me from years of medical training: Is the patient's airway clear, are they breathing, do they have a pulse? While Nate's head was bleeding and his leg clearly deformed, his screams afforded an odd sense of relief: he was breathing, conscious, and coherent while expressing his pain.

Nate and I were transported by ambulance to UCSF Benioff Children's Hospital in Oakland, where he was admitted to the pediatric intensive care unit (ICU) to await surgery on his broken femur the next morning. Nate had also sustained a skull fracture, so as a precaution, the ICU nurses checked on his neurologic status every two hours. They needn't have bothered. I kept vigil at his bedside the whole night, too wired by guilt to fall asleep.

The next morning, before 7:00 a.m., four members of the neurosurgery team came by on their rounds. The attending neurosurgeon started questioning Nate to assess his mental status: his name, his age, his grade in school. Then the surgeon glanced over at me. "Mom, I understand that you're a physician?"

Before I could speak, Nate interjected, "Yes, she is." Then out of nowhere, he added, "Hey, have you ever had herpes? Ask my mom—she knows all about it." I shook my head and closed my eyes, lowering my forehead into my hand. The team erupted in peals of laughter. The surgeon raised his eyebrows and looked at me. "Well, seems like he's clear neurologically."

Nate's accident happened to coincide with his realization a few weeks earlier of what I did for a living. Not just that I was a physician but one who happened to specialize in sexually transmitted infections (STIs). He had never talked about my job with others before this hospitalization, but he soon grasped that mentioning it would get a rise out of anyone. And he was going to milk it for as many laughs as he could.

During the hospital stay, Nate proceeded to chat with the ICU nurse about HIV, the orthopedic surgeon about syphilis, and to my chagrin, with the hospital chaplain about chlamydia. I would discover later that this behavior is a common phenomenon among the children of my colleagues. My boss's daughter wrote her college admissions essay about syphilis conversations over Shabbat dinner, while her son told his friends' parents that "my mom works in the sex industry." She's the chief of the Division of STD Prevention at the Centers for Disease Control and Prevention (CDC), but I suppose that's close enough.

As I watched my son's antics from his bedside, seeds of thought started to germinate in my mind. At the time of his accident in 2015, I had been working in the field of public health and STI research for eight years since completing my residency. STIs such as syphilis, gonorrhea, and chlamydia had been on the rise since 2000. Infections such as human papillomavirus (HPV) were so common, nearly every sexually active person would be infected at some point during their lives.

But even with the ubiquity of STIs, I knew that most people (even health care providers) simply don't feel comfortable discussing them. For most of us, *having* sex is much easier than talking about sex, especially its least pleasant consequences. Yet my son and colleagues' children had no problem talking about sex and STIs. They had become comfortable with the topic before they were old enough to realize that it was an uncomfortable subject for others.

During hours of downtime at the hospital with my son, I began to wonder if there was something I could do to make people more at ease discussing STIs. I knew I couldn't be overly ambitious. Some people have a hard-enough time discussing STIs with their sexual partners; I couldn't just expect them to bring up the topic with their mail carrier or local barista. Still, if I could spark a larger dialogue among the public about the topic, perhaps it would begin to reduce the stigma behind these ubiquitous infections.

By the time my son was discharged from the hospital four days later, I had a plan: I would write a book that would weave together storytelling, science, and humor to tell the little-known backstories behind various STIs. I dreamed that people would become so entertained by these tales that the ick factor around STIs might gradually be replaced with a bit of wonder and fascination.

Had the accident not happened, I would have done the sensible thing and waited until my kids left for college to write a book. I would not have started the process with one newly invalid kid and another still in diapers. Still, there is nothing like a little trauma to provide one with a foreshortened sense of the future. The accident, in addition to an earlier brush with death that I'd had during childbirth, made me wonder whether bad karma from a past life had caught up with me. I decided I'd better get writing before lightning could strike again.

I soon realized that tackling STI-related stigma would not be easy. The shroud of shame surrounding STIs is as old as the infections themselves. To many, STIs are considered to be "God's judgment for the sexually immoral and adulterous" (Hebrews 13:4) or punishment for fornication. And if STIs were a punishment, then by default, those afflicted with them should feel guilty. Never mind that STIs can afflict anyone, even someone who's only had sex within the confines of marriage.

Despite the stigma, I knew there were fascinating backstories behind my favorite sexually transmitted bugs that might capture people's interest. STIs have represented the interplay between sex and society as far back as the 1500s, when links between prostitution and disease outbreaks were first established. During the 1800s, prominent artists from Beethoven to Van Gogh suffered neurologic effects of syphilis that influenced their personalities and iconic works. More recently, STIs have played key but hidden roles in everything from World War II to the growth of the internet to *The Bachelor*.

STIs are also destined to remain part of our future. In 2019, the number of STIs reported by the CDC hit historic highs and are continuing to climb. The threat of multidrug resistance looms large for bacteria such as gonorrhea and *Mycoplasma genitalium*. There are viruses such as Ebola or Zika that we only recently discovered as STIs, which can linger in the semen for weeks or even months. Who knows what's next? We can't predict when another STI might emerge, but be assured—something *is* coming, and we'd better be ready when it does.

Strange Bedfellows is my attempt to explore the role of STIs in our past, present, and future. It is a journey that goes from the microscopic clashing of two human microbiomes during sex to the big picture of sexual networks, appreciating the influence that just a few players have over the sexual health of the many. We'll meander through the twists and turns of real people's sex lives and debunk common wisdom about STIs. You'll meet my dear colleagues, a scrappy bunch of scientists and public health workers who have chosen to devote their lives to this field. Then I've thrown in the details of my own journey: the tale of a studious Korean girl who ended up spending her days between other people's legs for a living. This book is a peek into my weird and wonderful world, and I hope you will love it here as much as I do.

Will shedding light on these hidden yet influential genital creatures help us defeat STI-related stigma? I don't know, but we must start somewhere. We've managed to defeat stigma surrounding previously taboo subjects such as cancer, creating discourse and shifting public sentiment toward support rather than shame. We need a similar sea change around STIs to have any hope of curbing the current epidemic.

Fortunately, reading about STIs presents no risk of actually catching one. However, I suggest you read the rest of this book with your clothes on. Otherwise, I make no guarantees.

Author's Note on Terminology: What's in a Name?

W hat's in a Name?" was the title of an editorial written by my colleague Hunter Handsfield in the journal *Sexually Transmitted Diseases* in 2015, which raised an important question: What term should we use to refer to the dozens of sexually transmitted bacteria, viruses, and parasites?

In the United States, the term *venereal disease,* or *VD*, was in vogue before the 1970s. It was a term that implied that venery, or immoral behavior, was involved in disease transmission. The 1970s ushered in the era of *sexually transmitted diseases (STDs)*, a term felt to be less stigmatizing than *VD*. We were happy with *STDs* for a while. But in the 1990s, there was growing recognition that some sexually transmitted bugs were silent infections that resolved spontaneously, meaning they didn't cause disease (e.g., HPV). Thus, the term *sexually transmitted infection*, or *STI*, came onto the scene.

Today, there still isn't consensus about which terminology should be used. While *VD* has fallen by the wayside, both *STD* and *STI* are still in use. Whether they're called *STDs* or *STIs*, you probably don't want either of them inside your pants. But if using *STI* helps reduce stigma, then I'm all for it. In fact, if something better eventually comes along, I'm prepared to throw *STI* to the curb in favor of another term.

All three of these terms are used throughout this book, because I wanted to preserve the language used by the people I interviewed or the names of organizations or programs, and reflect whatever term was in use at the time these stories take place. The subtitle of the book uses *STD*, as I felt that term would be most recognizable to both younger readers, generation Xers, and baby boomers. But I use *STI* as much as I can throughout the book, because that is where I think we are headed eventually.

Then we have *sex work* versus *prostitution*. We know that *prostitution* refers to sex work that involves exchange of sex for money. But sex workers can also be erotic massage therapists, exotic dancers, cam girls/guys, phone sex workers, and porn actors. I recognize that *sex work* is a more positive and more inclusive term for the diverse activities that go on in the industry. But when historical interviews or research refers to work with *prostitutes*, I've kept that language as it was originally referred to.

Words matter, particularly when it comes to topics as charged as sex and STIs. As I wrote the stories throughout this book, my aim was to maximize scientific accuracy, minimize stigma, plus educate and entertain along the way. I hope that I've achieved my goal.

1

Killing the Scarlet H

Stigma and Scandal in the World of Genital Herpes

Pray It Away

The woman sitting across from vaccine researcher Nick Van Wagoner at the University of Alabama–Birmingham was willing to do anything for a cure. She had been diagnosed with genital herpes in her midtwenties, and she had not had sex since her diagnosis. That was more than twenty years earlier.

A cure wasn't possible, Van Wagoner explained gently. Perhaps she could enroll in his clinical trial of a therapeutic vaccine; eventually, it could lead to a therapy that might lessen her symptoms. And such a vaccine might reduce the chances that she would transmit herpes simplex virus (HSV) to her future sex partners. Van Wagoner nodded as he said this, conveying the tacit message that yes, she would have sex again someday.

If she agreed to participate in the trial, she would be injected with the active vaccine or the placebo; Van Wagoner wouldn't know which she had received. She understood—she was ready to enroll. After she signed the consent form and left, Van Wagoner found himself thinking about her for days afterward. He knew for most people, herpes was not debilitating. Other than occasional genital discomfort, there were no lasting physical consequences of the infection. Over time, outbreaks were milder and became more of a nuisance than anything else; sometimes they stopped altogether. However, Van Wagoner could never predict people's psychological responses. They could range from a shrug of acceptance to the dramatic response of the woman in his office, for whom the shame of the diagnosis had upended her life.

More than a decade later, Van Wagoner still has vivid memories of that first patient he enrolled in a herpes vaccine trial. After working with hundreds of patients, he would come to realize that while the woman's reaction to her diagnosis was unusual, it was not unique. Abstaining from sex was the way some patients initially coped with the diagnosis. It didn't do anything to the course of the infection, but it prevented one from having to disclose their HSV status and risk the sting of rejection. For a time, one could live their life as if the diagnosis hadn't happened. But living in denial and avoiding sex was likely to fail eventually. Van Wagoner should know. He had tried to live that way himself for years.

Van Wagoner was born and raised in Utah, the youngest in a devout Mormon family. He first realized that he was attracted to other boys at the age of four. Even at that early age, he understood that it wasn't okay. "I quickly learned that such attractions were not only unwelcome but considered by God second only to murder in spiritual condemnation."

In the fifth grade, Van Wagoner first learned that HIV/AIDS was killing gay men, and his initial thought was, *That's what is going to happen to me.* Family friends' reactions toward the epidemic didn't help matters. "Either go to hell now or later," they remarked. Van Wagoner knew his fate was set. God would punish him for his attraction to men by giving him HIV, and he would die.

Terrified, Van Wagoner spent much of his teens begging God to change him. He tried to "pray away the gay." He kept up appearances by dating Mormon girls. One benefit of dating within his faith: he wasn't pressured to have sex. Then on his twenty-second birthday, his parents took him out to dinner. Afterward, they sat him down.

"Nick, we know you're gay."

While the shock of being outed by one's parents might cause other people to curse, Van Wagoner's inner Mormon held steadfast.

"Well, holy cow, how did you know?"

After the news of his sexual orientation spread through the community, church officials summoned Van Wagoner to a series of meetings. They reassured him, "Nick, it's okay. You can be gay, but you should never act on it." But given Mormon teachings, Van Wagoner knew that he could never reach the highest level of heaven without getting married. His choices were limited to "changing my same-sex attraction and remaining a member of my faith, or continuing on my current trajectory and suffering the spiritual and social consequences."

When it became clear that he was unlikely to change, Van Wagoner's church ward convened a council to exact their judgment upon him at a formal hearing. He knew there was only one possible outcome, but he attended the meeting anyway, mostly for his parents' sake.

The decision delivered was what he feared but expected: Van Wagoner would be *disfellowshipped*, a term used by the Latter-Day Saints to represent a type of excommunication. He lost his church community and most of his friends. But he left Utah and walked into his first day of graduate school at the University of

Alabama–Birmingham with a clean slate. He fell in love with his husband, Jeff, shortly thereafter.

After enduring so much stigma himself, it's fitting that Van Wagoner would devote his career to the study of HSV, one of the most stigmatized of the STIs. It was during graduate school that he first became interested in the inner workings of HSV, in particular the way the virus sets up shop inside people's bodies—a stealthy process with lasting consequences.

During sex, friction and microscopic trauma would allow HSV to enter the skin. Next, the virus would hijack cells' reproduction machinery, and these hijacked cells would begin mass-producing the virus until they burst open and died, releasing a flood of virus to infect other surrounding cells. Then the cycle would repeat. Once enough cells had died, an open sore could form at the original point of entry. For most people with HSV, this process was so subtle that they might not even realize it was happening.

Meanwhile, HSV was also hiding itself to ensure its continued survival. Instead of causing cell death and destruction, some particles of virus would travel quietly from the nerve endings in the skin to bundles of nerves at the base of the spinal cord. There, the virus could lie dormant for months or even years before it reactivated, traveling back out to the skin along the same path of nerves it had used to arrive, resulting in repeated outbreaks of genital herpes.

Of course, a person's immune system would respond, producing antibodies and sending inflammatory cells to try to kill the virus. But HSV was sneaky. With its ability to hide inside the nerve cells, it could never be eradicated—an unwelcome guest that would never leave.

Despite Van Wagoner's fascination with the physiology of HSV, what sustained his interest was the profound emotional impact of the infection. At the time he was finishing medical training in 2007, almost one in five Americans—an estimated 48.5 million people—had HSV-2, the type that caused recurrent genital herpes. Yet many of his patients felt like they were suffering alone. After he would lecture on herpes to the medical students, inevitably someone would come to his office and share how the diagnosis had affected their lives. Many of them broke down in tears. It was a genital stigmata, both seen and unseen.

It turns out that the stigma around genital herpes is much more recent than the infection itself. Joel Wertheim at the University of California–San Diego used molecular evolution models to figure out just how long HSV has been infecting humans and our ancestors. Lest you think humans are unique in their suffering, several primates—including macaques, chimpanzees, and baboons—have their own strains of HSV. But we are the lone primate unlucky enough to be afflicted with two different HSV strains.

For HSV-1, the culprit of most cold sores and fever blisters, Wertheim estimated that the virus genetically diverged from the chimpanzee herpes virus about 6 million years ago.[1] For HSV-2, associated with recurrent genital herpes, his models suggested that cross-species hanky-panky may have occurred between an ancestor of the chimpanzee and our ancestor *Homo erectus* 1.6 million years ago, giving rise to the new strain. Regardless of exactly when each strain began infecting human beings, both HSV strains were hearty enough to be passed down through generations of our various *Homo* ancestors to the modern-day man and woman.

But while *Homo sapiens* have been living with both strains of HSV for two hundred thousand years, the first descriptions of genital herpes didn't appear in the scientific writings until 1736.[2] That year, French physician Jean Astruc wrote the first textbook on STIs, *De Morbis Veneris,* as France was emerging from a devastating syphilis outbreak that had swept through Europe during the sixteenth and seventeenth centuries. In response, the French government mandated that sex workers undergo routine medical surveillance. As a result, French physicians such as Astruc acquired a rich knowledge of genital maladies in all their varying presentations.

In Astruc's texts, he described the classic herpes sores, the people who might be at risk, and where sores might be found on the body, but he didn't give the condition a particular name. Over the next several decades, his colleagues obliged, with names that evolved from *L'Olophlyctide progeniale,* to *herpés phylcténoide,* to *les herpès génitaux.* I wish we'd held on to one of these French monikers, including its elegant pronunciation (er-pehz). Branding isn't everything, but I think we'd be better off if the affliction sounded more like an expensive scarf or purse.

For more than two hundred years after Jean Astruc's first description of herpes, it mostly flew under the radar, overshadowed in the medical and public health communities by STIs such as syphilis and gonorrhea. People were certainly getting herpes, but there wasn't much public discussion about it. In the bestselling manual, *Everything You Always Wanted to Know About Sex* (*But Were Afraid to Ask)* (1969), genital herpes isn't even mentioned in the chapter on venereal diseases. In a Canadian manual, the *VD Handbook* (1977), there are fourteen pages devoted to the topic of gonorrhea, and only two to genital herpes.

The lack of attention paid to herpes may have been due to the absence of treatment options for people with the diagnosis. The *VD Handbook* admitted "there is no antibiotic yet available that can kill the virus." The handbook suggested painkillers, numbing creams, and wet compresses to reduce painful symptoms. For the worst cases, they suggested x-ray exposure of the genitals or cancer chemotherapy drugs.

Had science remained stuck in 1977, millions of herpes sufferers might be stuck irradiating their genitals or using ineffective treatments. As luck would have it, researchers in a sleepy North Carolina industrial park would soon make a discovery that would revolutionize the treatment of herpes. But no one could have predicted the fallout that would come in its wake.

The Manufacturers?

In 1981, all appeared calm from the exterior of the Burroughs Wellcome headquarters, nestled within seven thousand acres of forested terrain in North Carolina's Research Triangle Park. The company had commissioned the iconic building from architect Paul Rudolph, who had obliged with a futuristic mothership befitting the up-and-coming pharmaceutical enterprise. From the outside, it resembled a postmodern honeycomb; inside, it was full of natural light, diagonal lines, and soaring ceilings. Rudolph noted, "Anticipation of growth and change is implicit in the [building's] concept."[3]

Yet within the building's walls, growth and change had been painfully slow to come. For most of the decade that they'd inhabited their headquarters, a drought had plagued the scientists and executives at Burroughs. For eight years, they had not successfully brought a new drug to the market. It wasn't for lack of trying. The process of drug discovery and development was complex and unpredictable; it remains this way today. Only one in ten drugs that makes it through human clinical trials actually reaches today's market, not to mention the thousands of candidates that first fail through the rigorous clinical trial evaluation process.[4]

By March of 1982, there was a glimmer of hope. Burroughs became the first pharmaceutical company to receive approval from the Food and Drug Administration (FDA) to market acyclovir ointment (Zovirax) for the treatment of genital herpes.[5] Acyclovir was the first antiviral drug of its kind, able to suppress the virus but not eradicate infection; it would later serve as a prototype for antiviral drug development against HIV.

At the time, there were no other drugs that treated genital herpes, yet the marketing department at Burroughs was not optimistic. According to Pedro Cuatrecasas, then head of research and development, most of the marketing team had never heard of genital herpes and therefore doubted that there would be a significant market for the drug.[6] They predicted sales of acyclovir would be modest, peaking at about $10 million per year. Although this may seem like a large windfall, it paled in comparison to so-called blockbuster drugs such as the antacid Zantac, which had annual sales of $2 billion during the same era.

Still, Burroughs's team would try to market what their scientists had developed. But how were they to generate interest in a condition that they had never heard of?

Time magazine unexpectedly lent a hand to Burroughs's cause. In August of 1982, they published a cover story titled "Today's Scarlet Letter," with a giant *H* and the word *Herpes* emblazoned across the cover in bloodred ink.[7] It was not the first time that *Time* had tried to generate interest on the topic. Two years earlier, they had published a piece entitled "Herpes: The New Sexual Leprosy." The message in both pieces was similar: that genital herpes would transform "nice, healthy, educated, clean-cut Caucasians of the upper and middle classes" into personae non gratae in the bedroom.

But the new story implied that herpes held a greater power. It was "the scourge" threatening to undo the sexual revolution of the 1960s, forcing Americans toward a "reluctant grudging" return to chastity. It warned swingers, philanderers, and prostitutes' johns that their behavior was a "high-stakes gamble" for catching herpes. The piece reduced herpes sufferers to "herpetics," as if their disease comprised their entire identity. There were interviews with these downtrodden herpetics describing themselves as "poisoned," "unmarriageable," and "depressed."

Time's article predated the emergence of AIDS, before anyone knew that a viral STI (i.e., HIV) could actually kill you. Still, the most punishing tales in the article were reminiscent of those from the early AIDS era: law firms wondering about the legality of firing employees with herpes, and coworkers petitioning to ban a woman with herpes from the office and refusing to share a phone with her. (Did they fear she would rub the receiver all over her vulva?)

The *Time* article never mentioned acyclovir, nor its manufacturer's name. But the public's growing preoccupation with herpes increased their interest in a treatment, and Burroughs's product was the only one on the market. The company's profile rose exponentially as a result. According to *The New York Times*, from 1982 to 1983, more than one thousand news articles were written about Burroughs Wellcome and acyclovir.[8] Still, simply raising the profile of the company and the product would not guarantee success. William Sullivan, president of Burroughs Wellcome, was circumspect about the company's newfound fame. "The time for the real hoopla," he said, "is when the product takes off."

The initial formulation of acyclovir ointment didn't turn out be a home run. It was only approved for an initial herpes outbreak, and it seemed only to reduce the pain from outbreaks in men. Luckily, Burroughs had something else up their sleeve. They had also formulated acyclovir into a capsule, taken orally, that could speed up healing of sores from an initial outbreak and recurrent outbreaks. If the FDA would approve their new formulation for recurrences, this could mean multiple sales of acyclovir per patient every year.

By the end of January 1985, Burroughs's wish was fulfilled. The FDA approved acyclovir capsules for both men and women with herpes, which could be used for initial and recurrent outbreaks. At the time, according to the CDC, there were more than thirty-four million people with herpes in the United States. Now Burroughs just had to get some of these patients and their physicians interested in their drug.

This is where direct-to-consumer marketing came in, but not quite in the way you are used to seeing today. Direct-to-consumer marketing often involves ads where the problem, the product's name, and its claim are presented together, sometimes in way that really hits you over the head. Here is one of my favorite examples:

> Gentlemen: VIAGRA helps guys with erectile dysfunction get and keep an erection.

What image does this evoke for you? I imagine a man, a blue pill, and an erect penis, which is exactly what Pfizer wants. In 2016, pharmaceutical companies spent an estimated $5.6 billion on direct-to-consumer ads like this for one simple reason: they work.[9]

But this was the 1980s, and direct-to-consumer advertising was still in its infancy. The FDA didn't explicitly forbid it, but they had called for a voluntary moratorium on the practice between 1983 and 1985.[10,11] According to Louis Morris, the FDA's division director for drug advertising, the agency had multiple concerns about marketing directly to consumers, which are still relevant today. They worried that it would "lead patients to pressure physicians to prescribe unnecessary or un-indicated drugs . . . potentiate the use of brand name products rather than cheaper but equivalent generic drugs, and foster increased drug taking in an already overmedicated society."[12]

By 1986, the FDA had lifted its moratorium, but Burroughs Wellcome still treaded lightly. They came up with a series of help-seeking ads, which made no mention of their product, or the product's claim, but tried to increase attention to the disease and encouraged consumers to seek medical attention.

One ad series featured a Caucasian couple named Roger and Sally at the ocean: Roger, tall and handsome, appeared contemplative as he wrapped his arm around Sally, a petite smiling brunette. Depending on the publication and the intended audience, the ads alternated which of them actually had herpes and had to face the task of disclosing their status to the other person. The copy for each ad was similar:

> *The hardest thing she ever had to do was tell Roger she had herpes. But thanks to her doctor, she could also tell him it's controllable.*

Not to discount the challenge of disclosing one's herpes status, but if I were Roger, I can imagine dozens of things that would be harder to hear about than Sally's herpes. How about: "I just gambled away our savings," or "I'm sleeping with your brother."

Whether intentional or not, the message behind the ad was clear: 1) herpes is shameful and it would be excruciating to disclose it to one's partner; 2) one should run and seek medical attention because there is now a prescription that would make life easier. There was no mention of the name *acyclovir,* and the company name and logo were nondescript, printed in tiny font at the bottom of each ad.

Despite the subtlety of the ads, acyclovir capsules were a smash hit. Shattering all prior predictions, sales of acyclovir reached over $1 billion annually, eventually accounting for a third of Burroughs's sales and half of its operating profits. One of Burroughs's pharmacologists, Gertrude Elion, was awarded the Nobel Prize in 1988, due in part to her discovery of acyclovir. The company's success with acyclovir caught the attention of pharmaceutical giant Glaxo, who would acquire the company in 1995 for $14 billion.[13] Not bad for a drug treating a condition few had heard of just fifteen years prior.

Burroughs certainly deserved to reap profits from their hard-earned discovery. Acyclovir is a medication that physicians refer to as a *good drug*. Although it doesn't actually eradicate the virus, it's safe, reduces or eliminates symptoms, and has no significant side effects, even among patients who take it continuously for years. For more than two decades, it has remained on the World Health Organization's Model Lists of Essential Medicines, used by governments in 155 countries to determine which medicines to prioritize for public health.

However, some journalists see a dark side to acyclovir's success, accusing Burroughs of manufacturing herpes stigma to increase their profits. But if Burroughs's 1986 ad campaign stoked the fire, they certainly didn't light the match—*Time* magazine's "sexual leprosy" and "scarlet letter" messaging in 1980 and '82 likely played a greater role of raising herpes consciousness and stigma. Still, it seems that herpes stigma had been present long before acyclovir and *Time*'s articles came on the scene.

Lawrence Corey, a pioneering herpes researcher from the University of Washington, describes this notion of Burroughs manufacturing stigma as "totally fallacious." He worked alongside Burroughs scientists to develop acyclovir, and in the pre-acyclovir days he ran a specialty clinic for genital herpes at the university. At the time, he could offer little to help his patients. He tried topical formulations of several anticancer drugs without success. He resorted to applying ether to herpes lesions, causing some of his patients to cry from pain.[14] Some told Corey that his treatments were worse than the disease.

"People were really devastated by this disease," he recalled. "It was very stigmatizing because there was no therapy. Women and men were coming to me to participate in these [treatment] studies. They would get up at six o'clock in the morning to catch the ferry to do this. They were dying for something to happen, something that would help them handle this sexually transmitted infection, which had been ignored or trivialized. The stigma around HSV has always been there. It's always been an issue."

Corey agrees that the public and the medical community largely ignored genital herpes before acyclovir arrived on the scene. And he still thinks the medical community gives herpes the short shrift—it's not deadly enough to get the same respect as other STIs like HIV. But the emergence of effective therapy finally caused people to pay herpes some attention.

"Having a therapy and having a tool meant that doctors had to learn about it. A lot of the awareness came about because you had the antiviral," he said. Corey did his part to raise awareness by publishing numerous landmark studies on the natural history of the virus and the clinical presentations of genital herpes that he saw in his clinic, allowing American scientists to reacquaint themselves with *les herpès génitaux* that the French had first described in the 1700s.

It also helped that improvements in testing began to allow for many more diagnoses to be made. In the early 1980s, a diagnosis of herpes could only be confirmed with a viral culture or other techniques to detect HSV-2 from open sores. Thus, only patients with symptoms could be certain of the diagnosis. This would change by the end of the decade when the first herpes antibody tests that could distinguish between HSV-1 and HSV-2 hit the market. Finally, anyone in the public could request a blood test for HSV-2, regardless of symptoms.

It turned out that people with herpes symptoms were just the tip of the iceberg. By 1994, the CDC estimated that forty-five million Americans (one in five) had HSV-2.[15] Most of these people were walking around unaware, yet they were capable of transmitting the virus to others. Millions of Americans flocked to be tested, to find out whether they were part of that one-in-five statistic. On the surface, this seemed like a good idea. Everyone should know their status so they could use condoms and acyclovir to reduce transmission. Knowing was always better than not knowing, wasn't it?

The Bachelor

Shawn was feeling a glimmer of excitement to be single again. He had just returned to Washington State after his final overseas deployment with the army. After years of struggling in a long-distance marriage, he and his wife finally agreed to call it quits. He decided to try his hand at online dating and met Eva, who was dealing with a similar relationship situation. They navigated the process of undoing old ties with their exes and creating new ones with each other.

For the first month, their relationship was limited to talking and kissing. Before going further sexually, Eva wanted them to both test for STIs and share their results with each other. She had gotten into the habit of doing this frequently as her own marriage was ending, when she and her husband tried polyamory in a last-ditch effort to save their relationship.

No one had ever asked Shawn to test for STIs before, but he was game. He'd spent more than a decade in the army and had slept around more than Eva, who married the first man she'd ever had sex with. And Shawn was informed about STIs—the army had made sure of that. Early in his service, his commanding officers summoned him and his fellow soldiers to a darkened theater and forced them to watch extreme cases of herpes and genital warts on the big screen. While the army was obviously concerned about STIs, Shawn and the members of his company were much more afraid that they'd get someone pregnant and have to deal with the aftermath. Despite the army's scare tactics, their fear of STIs was minuscule when put up against hormones, alcohol, and willing partners overseas.

Shawn made an appointment at a local clinic and asked the clinician to test him for STIs. About a week later, he approached the front desk to pick up a copy of his results to show to Eva.

In a breach of standard protocol, the receptionist perused the printout before handing it over to Shawn, saying, "It looks like you're okay except for this one."

"Well, what is that?" Shawn gestured toward an abnormal result on his report.

"Oh, that? That's herpes."

The receptionist didn't ask him whether he'd like to speak to a doctor or nurse, so he calmly took his results and left.

For a moment, Shawn wished that he were back in the army. The army had standard procedures for delivering bad news. No one would ever just drop a bombshell on someone without some follow-through.

But now that Shawn was a civilian, he had none of those procedures to protect him. He sighed as he looked down at his results again. He had never experienced symptoms of genital herpes, but clearly his antibody test was positive for HSV-2. *Of course,* he thought, kicking himself. *It's probably something stupid I did when I was in the army. I've finally met someone great, and now I have to tell her that I have an incurable STI.* He wondered how many other people he'd have to go back and tell.

That evening, Eva was quiet after Shawn showed her his results. For several days, they both digested the information and didn't speak about it. When they spoke again, they agreed that if their relationship were going to work, they'd have to figure out how to handle it.

Although the rest of Shawn's STI results were negative, they hadn't tested him for another common STI, human papillomavirus (HPV). HPV wasn't related to herpes, but Eva figured that if Shawn had one viral STI, he could have another one. At least she could get a vaccine against several strains of HPV. The series of three vaccinations would take her six months to complete. They agreed to keep it platonic for the time being and figure out how to navigate their sex life once those six months had passed. In the meantime, Shawn promised that he'd look into treatment and management options for herpes.

Shawn had interrupted college during his military service, so he went back to school on the GI Bill. Through his university's library, he began fact-checking what he had already read online about genital herpes, poring over original research in scientific journals. Each study he read would reference twenty other studies, so he'd read those too. It was like staring at something in parallel mirrors, seeing ever-smaller images tunneling into infinite distance. For months, he dove into these tunnels, studying herpes vaccines, diagnostic testing, clinical symptoms, psychological effects, transmission, treatment, viral suppression. As his six-month waiting period with Eva was ending, he'd learned a lot but was still unclear how to put it all together.

During his research, Shawn noticed that all paths kept leading to the same person: Anna Wald, a physician and researcher at the University of Washington. He looked up her online profile, and right below her picture was her research clinic's address and her email. *What the hell*, Shawn thought, and he decided to write to her. He spelled out his sexual history and his test results to see what advice she might have for him.

After more than three decades of herpes research, Wald was used to getting questions like Shawn's. She always felt obliged to respond. "I'm a public university employee; my salary comes from taxpayer dollars. And all taxpayers are at risk for herpes." She invited Shawn to visit her research clinic if he wanted to confirm his original test results with a Western blot, the same type of test used for decades to confirm antibody results for HIV. But unlike HIV, the herpes Western blot was only available in one place, the University of Washington's virology laboratory.

Shawn lived just half an hour south of Wald's office, so he took her up on the offer. The next week, she called him and left a cryptic voice mail. Shawn figured she was concerned about his privacy. He emailed her to reassure her that it was okay to leave him a message with his results. "Look, it's fine, I already know I have it," he wrote.

"No," Wald replied. "I need to talk to you."

When they met, before discussing Shawn's results, they reviewed the Western blot of a patient with HSV-2. On two long, narrow strips of paper, Shawn could see more than a dozen dark striped bands, each band demonstrating an antibody to a different protein of HSV-2.

Then he looked at the image of his own Western blot. He saw a single band: it looked nothing like the other patient's results.

Before saying anything, Wald probed a little more about his sexual history and whether he'd had any new sex partners. Was it possible that he had just acquired a new herpes infection? If so, he'd develop more bands over the coming weeks. Shawn shook his head. He hadn't done anything with anyone in six months. He had been waiting for Eva.

That left Wald with one other logical explanation: that Shawn's original blood test had been falsely positive. Shawn wondered if he was perhaps exposed to something on his deployments to the Middle East that triggered the positive result. Wald couldn't say. But she was confident that he didn't have HSV-2. Shawn had just spent the last six months trying to cope with something that didn't exist.

Shawn couldn't believe his luck. He didn't have HSV-2, and Eva had stuck around even when she thought that he did. During his six-month wait, Shawn had decided that if things didn't work out with Eva, he'd try online dating on Positive Singles, or one of the other dating sites for people with HSV-2. If Eva had left and had he not reached out to Anna Wald, he would have sought out sex partners with HSV-2, possibly inadvertently infecting himself in the process.

Unfortunately, Shawn's situation is not unique. The most commonly used HSV-2 antibody test in the United States is quite sensitive and is able to detect even low levels of antibodies to the virus (so-called low positive results). Any numerical value above 1.0 is considered positive, but Wald has found that low positive results like Shawn's (between 1.1 and 3.5) can often be false positives.

False positive tests are so problematic that U.S. guidelines updated in 2016 recommended *against* testing the general public without symptoms for HSV-2 with these antibody tests. In the analysis accompanying the recommendations, the United States Preventive Services Task Force estimated that the HSV-2 antibody test could give false positive results as much as half the time, particularly if the results were in the low-positive range.[16]

In 2017, Wald investigated this for herself. She and her team reviewed test results of patients who'd been tested with a standard antibody for herpes and also with the Western blot from the University of Washington. It was worse than she'd feared. Of 381 people with a positive antibody test for HSV-2, only half were confirmed with the Western blot. The other half were false positives, which were more likely in people who had cold sores or antibodies to HSV-1 in their blood.[17] Wald reflected, "These tests aren't as good as they ought to be, given that they are used to diagnose someone with a chronic, lifelong, sexually transmitted disease."

To add insult to injury, false positive HSV-2 tests may be keeping men and women from finding true love, at least on reality TV. According to Amy Kaufman's tell-all book *Bachelor Nation*, genital herpes is the most common reason that potential contestants are rejected from the show. When the production team becomes aware of the test results, the showrunner's assistant breaks the news to the contestant that they can't participate, accompanied by a vague statement along the lines of, "You should call your doctor."[18]

Do *The Bachelor*'s producers realize that many of these herpes tests could be wrong? Imagine all the amazing women who have been turned away on false pretenses. And if this is truly "reality" TV, why not include women with HSV-2 on the show and let the Bachelor decide what to do? Imagine what could happen to herpes stigma if millions of people saw this:

Bachelorette: "I'm HSV-2 positive; I have the virus that causes genital herpes."

Bachelor: "That's okay, honey. We'll figure it out." (Kneels and slips 3.5-carat Neil Lane diamond on finger. Fade in romantic orchestral music.)

To all the rejected *Bachelor* hopefuls who were told they have HSV-2, it might not be true. Get yourself a Western blot before giving up on that rose.

Luckily, true love is greater than the threat of herpes. Shawn and Eva eventually married; according to Shawn, they are living happily ever after. Anna Wald retold their story during an awards acceptance speech she gave in 2015: "The threat of HSV-2 discordance led them to increase their intimacy and to a decision that they will remain involved despite the belief that he has HSV-2, and she does not."[19] She appreciated Shawn's attitude toward the false diagnosis: "The herpes diagnosis served a useful purpose," he wrote, "and we're happy for the effects it had on us, but . . . we won't miss my 'pseudo-herpes.'"

It's easy to see the silver lining in Shawn's situation. He went through six months with a diagnosis of HSV-2, magically undone by a negative Western blot. He found true love in the process. But there are certainly others who aren't as lucky, who never got the chance at a Western blot and may be taking antiviral medication unnecessarily.

And what about the millions of Americans who actually do have HSV? For patients who suffer from frequent symptoms, some develop a sense of urgency, even desperation, to find a cure. Herpes vaccine researchers can relate; they've spent decades testing vaccines and new antivirals for HSV and have come up empty-handed. Then one summer in 2016, the collective desperation between a group of patients and a vaccine researcher came to a head, driving them all right over an ethical cliff.

Desperate Measures

A warm breeze washed over Richard Mancuso as he descended the stairs onto the tarmac at Saint Kitts's airport, filling him with hope that his luck was about to change. Since the early 1990s, he had endured extensive herpes outbreaks over his groin and his face, which had crippled his social life and his self-esteem. Antiviral drugs had helped temporarily, but they also caused severe headaches, so he could never stay on them for long. When he took a break from antiviral drugs, he would have outbreaks as often as two or three times a month. At times, he became depressed, even suicidal over his situation.[20]

After posting about one of his outbreaks on a Facebook page for people with herpes, the site's moderator connected him to William Halford, a microbiology professor from Southern Illinois University (SIU) who had developed a new therapeutic vaccine for HSV. Halford had already spent ten years tinkering and testing different versions of his vaccine on animals. He felt the moment had arrived to conduct a trial in humans.

Halford's vaccine was based on live strains of the HSV-2 virus, which he'd genetically engineered by deleting pieces of DNA from the viral genome. He believed a live vaccine would be superior to the previously tested *subunit* vaccines. These subunit vaccines were based on proteins from the surface of the virus and were thought to be safer than a live vaccine. But several iterations of subunit vaccines had already been tested and failed.[21] In Halford's estimation, these vaccines couldn't sufficiently protect people from contracting HSV, nor effectively diminish recurrences for people already living with herpes.

Despite his belief that his vaccine was effective, Halford knew it would take another ten years to go through clinical studies and the FDA's arduous approval process to bring the vaccine to market. He also knew that he didn't have that kind of time. Sitting inside his face was a golf ball–size tumor, jutting through the bony cribriform plate that separated his sinuses from the floor of the skull and frontal lobes of his brain. His oncologist had confirmed that it was sinonasal undifferentiated carcinoma, a rare cancer with a five-year survival of less than 30 percent.[22]

To avoid jumping through the FDA's regulatory hoops for manufacturing a vaccine and conducting a Phase I clinical trial, Halford decided to conduct his trial on the Caribbean nation of Saint Kitts and Nevis. The offshore trial would be conducted in secret. He did not seek approval to conduct research on human subjects from his university's institutional review board or the FDA. And the local government had no idea that he planned to inject people with this new vaccine in private homes on the islands.

Before testing his vaccine on others, Halford first injected himself and three members of his immediate family with his vaccine, which he named *Theravax*$^{HSV-2}$. None of them seemed to suffer any ill effects. He touted these results as evidence of his vaccine's safety. In a documentary about his work, he reasoned, "If we didn't have a problem, you shouldn't have a problem." Encouraged by the results of his self-experimentation, Halford began recruiting men and women with herpes to the Saint Kitts trial.[23]

On Halford's recruitment call with Mancuso, he spent two hours explaining the study, answering Mancuso's questions, and outlining the risks and benefits of participation. Mancuso was impressed by Halford's demeanor. "He was a very calm and soft-spoken man . . . At times he spoke simply and bluntly with a matter-of-fact attitude—not in the negative sense, but in the sense that you would get from someone who really knows the material."

Halford's trial consent form was similarly blunt: "*Theravax*$^{HSV-2}$ vaccine is not an FDA-approved treatment, nor has the *Theravax*$^{HSV-2}$ vaccine been manufactured per the Good Manufacturing Practices criteria required to initiate an Investigational New Drug application with the FDA." Halford acknowledged in the consent form that the FDA played a vital role in protecting patient safety, but he also estimated that one hundred million people worldwide would contract HSV-2 during the time that it would take to go through the FDA's standard regulatory processes. This was simply unacceptable to him. Near the end of the form, Halford promised that he and his newly formed biotechnology company, Rational Vaccines, would make his line of HSV vaccines available by 2017.[24]

Mancuso felt torn between his options. On one hand, if he did nothing, he'd continue with severe and frequent herpes outbreaks. On the other hand, flying to the Caribbean for an experimental vaccine under unorthodox conditions seemed risky.

He reflected, "I debated heavily: 'I'm taking a big chance here,' I said to myself. 'But if this is the real deal, it'll end up being my personal cure.' And after 20 plus years of suffering, that was a chance I was willing to take. I will say, if I'm being completely honest, I almost backed out."[20]

Mancuso flew to Saint Kitts three times to complete the series of injections with $Theravax^{HSV-2}$ from April to July of 2016. After his first shot, he developed intense redness and inflammation at the injection site, along with a fever and flu-like symptoms. His outbreaks continued, but as he charted them on a calendar, he noticed they were smaller in size, frequency, and duration. Less than a year after he began the trial, his outbreaks stopped altogether.

In an October 2016 press release, Halford's company, Rational Vaccines, declared the Saint Kitts trial an overwhelming success.[25] They claimed that the seventeen participants (including Mancuso), who received three doses of $Theravax^{HSV-2}$, all reported a reduction in symptoms. On average, participants reported a threefold reduction in the number of days they had symptoms, compared to their prior experience on antiviral drugs. Some, like Mancuso, were "functionally cured," or completely free from outbreaks.

Halford attempted to publish the results of the Saint Kitts

trial in December 2016. His eccentric manuscript blended personal history, philosophy, results of his self-experimentation, animal studies, and trial data from Saint Kitts.[26] It was rejected by the journal *Future Virology* and has not yet been published. Lack of publication notwithstanding, the trial results were enough to garner funding from PayPal cofounder and Trump adviser Peter Thiel, bringing the total investment in *Theravax^{HSV-2}* to $7 million by April 2017, just two months shy of Halford's death.

Contemporaries who learned of Halford's rogue tactics were horrified when they realized he'd tested a live vaccine on patients without oversight from a human subjects' committee or the FDA. "What they're doing is patently unethical," said Jonathan Zenilman, chief of Johns Hopkins Bayview Medical Center's Center for Infectious Diseases. "There's a reason why researchers rely on these protections. People can die."

One of Zenilman's colleagues contacted Marisa Taylor of Kaiser Health News, who broke the *Theravax* story in August of 2017. The series of stories that followed exposed Halford's unauthorized trials, both the Saint Kitts study and another that he'd conducted at a local Holiday Inn Express and Crowne Plaza Hotel three years before. Taylor's exposé led to an internal investigation and halt to all herpes research at SIU, a letter from the Senate Judiciary Committee to SIU demanding corrective action, and a criminal investigation into Halford's work by the FDA.[27–31]

Mancuso realized the optics of Halford's conduct were unfavorable. In his memoir *Asking for a Friend*, he reflected on the fallout of Halford's studies:

> They [the media] have used this information as a *hit piece*, to say the trial was unethical and to also imply that some participants in the trial were taken advantage of. To be

honest, if I had observed from the outside, rather than having been directly involved, I'm sure I could have easily assumed many different conclusions and had many similar questions. Despite the rumors, I never saw anyone taken advantage of or mistreated in any way. In my opinion, I am also confident that, in time, the facts and science will prevail.[20]

It may be a while before that science sees the light of day. Rational Vaccines' Facebook and Twitter pages went silent on August 2017, their website just went live again in July of 2020. In April 2018, the company's CEO, Agustin Fernandez, commented to CNN on its future plans for *Theravax*[HSV-2]. They were "currently in the process of moving forward with classical clinical development plans, including engaging established contract research organizations, contract manufacturing organizations, filing worldwide [Investigational New Drug applications]—focused initially in the U.S.—and performing clinical trials to international good clinical practice standards."

Even if Rational Vaccines is not criminally liable for Halford's actions, they will still need to deal with a lawsuit from three other study participants, claiming they suffered adverse side effects following vaccination.[32] Whether Rational Vaccines and *Theravax*[HSV-2] will emerge from the federal investigation and lawsuit remains to be seen. If they do undertake a new vaccine trial, surely there will be droves of herpes sufferers who will volunteer, regardless of the risk. Linda Oseso at the University of Washington found that over 40 percent of patients with HSV-2 were willing to endure severe side effects or hospitalization for an experimental treatment if there was possibility of a cure.[33] HSV possesses this singular power: it motivates ordinary people to do extraordinary things in the hopes of sparing themselves and others from further pain and stigma.

A Little Better All the Time

Emilia's tears welled up on the walk from the waiting area to the exam room and began to spill over as soon as I shut the door. Searching in vain for tissues, I handed her the softest thing I could find, a few squares of gauze, which she used to wipe her eyes. She had been here just a week earlier, she explained. She had been feeling fatigued for several days and her muscles ached all over. She assumed she was coming down with the flu, until a crop of painful blisters appeared around her vagina.

One of the nurse practitioners had gently unroofed one of the blisters, collecting the clear fluid onto a swab that she placed in a tube of pink liquid media. The swab was tested by polymerase chain reaction (PCR), the same technique used at crime scenes to detect DNA from blood or other body fluids. PCR testing for HSV has been commercially available since 2010 and is more sensitive for detecting HSV than traditional culture methods.

Between clinic visits Emilia had pored over photographs of herpes online. She'd interrogated her boyfriend and painstakingly inspected his genitals. Today, she'd brought him with her so that we could do the same. Emilia figured that she had herpes but was holding out hope that she would have HSV-1. Genital HSV-1 tends to cause a single outbreak, then sits quietly and lets you go about your business. Genital HSV-2 tends to recur, but the frequency and intensity of outbreaks for a given person is hard to predict.

Glancing at her results, I saw that Emilia's PCR test was positive for HSV-2. Unlike the antibody test, false positives were not an issue with PCR, which is performed directly on a genital lesion. Her test results, coupled with her symptoms, made the diagnosis a certainty. She shook her head and closed her eyes, resigning herself to this reality.

I told her that symptoms should improve over time. This first year or two would likely be the hardest. She could take acyclovir continuously to reduce her chances of having an outbreak and decrease the risk of transmission to sexual partners. I continued talking until I realized that Emilia had stopped listening. What if things didn't work out with her current boyfriend? Emilia asked. How would she explain this to someone new? She had already moved beyond the previous week's symptoms, her current state, even her current relationship. She was thinking about herself at some unknown moment in the future.

According to Anna Wald, Emilia's response was typical. She was trying to figure out how to deal with this new version of herself, a person now living with herpes. Wald observes, "Diagnosis of genital herpes throws you off the path of your narrative about yourself and requires the hard work of making a new one. How far it will throw you, how hard you land, which direction you choose afterward will depend on your personality construct, psychological resources, disease severity, and social support."[19] Wald receives numerous emails from people overwhelmed by a sense of loss at their diagnosis: "My life is over." "Completely lost." "I can't sleep, I cry all the time because someone stole my dreams from me."

If the sense of loss persists, an intense desire to clear HSV from the body can arise. Snake oil dealers capitalize on this anxiety, offering thousands of alternative remedies that promise a cure. Peddlers tout their herbal products on YouTube: oregano oil, thyme, red algae. There are those who claim dietary manipulation is the key: consuming raw food, avoiding foods high in arginine (think nuts and seeds). While I'm all for clean eating and the healing power of food, no diet or product eradicates HSV from the body. Any improvement in symptoms is difficult to tease apart from the natural course of herpes, which is that it usually gets better with time.

On another end of the coping spectrum, there are those who turn their diagnosis into an opportunity to help others. Christine Johnston at the University of Washington spends a lot of time with these folks—let's call them *herpes altruists*—those willing to volunteer themselves and their genitals to enhance our understanding of HSV. Johnston is an expert on viral shedding, the process whereby HSV comes off the surface of the skin, rendering a person infectious to others, and the success of her research depends on such herpes altruists.

Viral shedding is a sneaky phenomenon. It happens imperceptibly, and it can be rare and random at times, and continuous and copious during an outbreak. It diminishes over time, from a third of the days in the first year, to about once a week after ten years.[34,35] Sexual transmission can readily occur during times of silent shedding, and there are no drugs available to suppress it fully.

To tease apart the phenomenon of viral shedding—when, where, and how often it happens—Johnston must find herpes altruists with an intense commitment to the cause. In one of her studies, participants swabbed their genitals all over with a long Q-tip, four times a day, for up to two months. They kept a daily diary of symptoms, recorded each time they had sex, and visited Johnston and her team every two weeks to check in.

During this study, Johnston decided to map the exact location on the genitals that shedding was occurring on a given day of the month. She created maps of the male and female genitalia, dividing each into distinct regions with a grid. She recruited patients to have each mini-region of their genitals tested for shedding by PCR: twenty-two swabs a day for women, twenty-six for men, five days a week, for an entire month. Each patient provided more than four hundred swabs during the study. At the end of the study, patients received genital biopsies from areas that appeared to be shedding HSV.[36]

Johnston translated these results into heat maps representing hot spots of viral shedding over time. She was able to demonstrate that shedding was widespread on both sides of the genital tract regardless of symptoms, even though outbreaks tend to recur just on one side.

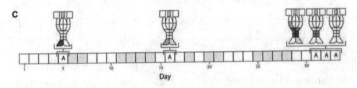

Heat maps of the vulva and perineum showing the location, frequency, and intensity of viral shedding in a single patient over 30 days. The letter A represents days of silent (asymptomatic) shedding. From Johnston, et al. *Journal of Virology*, 2014.

For Johnston's participants, their perseverance with her study protocols have nothing to do with financial gain. Compensation for her studies was minimal, far less than one could make volunteering in drug trials for pharmaceutical companies. Johnston's theory is that participants' dedication and altruism is a coping mechanism to deal with the negativity surrounding their diagnosis: "I think it is all tied into the stigma, that they felt so stigmatized when they were diagnosed, or they have such a difficult time with the way some people reacted. They want to give back. They want to improve the situation for the next generation."

Johnston sees her study participants becoming empowered as they learn more about HSV. "I feel like they do get something really positive from being involved, that they're both giving something back, and also getting something from being in the study," she said. "They are so committed to making this better."

Aside from becoming a guinea pig for science, most people find a way to cope with their diagnosis and move on with their lives. For Emilia, just diagnosed with HSV-2, the wound was still fresh as we sat together in the exam room; it was difficult for her to see beyond it. I longed for her to meet her future self, to see how time and experience could offer perspective. Her feelings might evolve to be more like Gail's, who had been living with herpes for more than twenty years. Like Emilia, Gail had contracted herpes from a partner who had no symptoms, and she had suffered debilitating outbreaks until going on antiviral medication. Now she was newly single after a recent divorce, entering the world of online dating, and tackling the disclosure of herpes to new partners.

I sat with Gail at her dining room table, where we pored over her OkCupid questionnaire. We read over hundreds of questions the company would use to cross-reference with a potential match. They ranged from scientific views—"Is climate change real?"—to hypothetical sex questions: "If a clone was made of you, would you sleep with it?" There was also a question about whether she'd consider sex with someone with herpes.

Gail knew friends who had posted their herpes status boldly on their profile pages along with their pictures. She wasn't ready to go that far but didn't feel that she had anything to hide. She had initiated disclosing her status with one guy because she wanted to move things along sexually. She explained that she was taking daily medication to reduce her chances of transmission. Her partner seemed unfazed. The next guy she dated beat her to the punch.

"He asked me straight out, 'What's your STI status?' And I was a little taken aback, but not entirely. And I told him at that point, I said I had genital herpes, and he said, 'I do too.' I felt a lot of relief. There were a lot of things I liked about this guy. Now I can check that box, like that works too." She checked off an imaginary list from her dating profile: "You know, the arts, the outdoors, herpes."

Ultimately, Gail felt that herpes was just one more thing that had to align to make a relationship work. "Going along with the general thing of dating and trusting, some people will work out, and some people won't. Whether the other person has herpes too and it's a non-issue, or they're someone who can work through discomfort or challenge their own assumptions . . . I see herpes more in that context . . . It's just one of a whole number of things that have to work out . . . I can see that it [disclosing her status] will get easier over time."

Perusing OkCupid's other questions for her profile, Gail mused, "The way things are looking, I think I'm going to find more people who have trouble with the fact that I don't shave off my pubic hair than with the fact that I have herpes . . . or don't want to slap a guy's face to have an orgasm, or squeal like a dolphin during sex."

It didn't occur to Gail to limit her online dating to sites for people with herpes: Positive Singles, Meet People with Herpes, H-Date, and my personal favorite, STD Soulmates. While the internet is a great place to find supportive community, I don't think people with herpes should be limited in who they can date. For herpes stigma to get any better, we need to keep mixing people up and having conversations about herpes, one person at a time. I guarantee every online dating site has people with herpes, whether people realize their status or not—Tinder, Match, Grindr, and yes, even Christian Mingle. Besides, changes are brewing that could make these herpes dating sites obsolete: people with herpes may soon be harder to find.

Extreme Makeover

If you are looking to feel inadequate as a mother, I would suggest taking your child to the Upper Noe Recreation Center in San Francisco. There you can compare yourself to moms who are carrying $300 designer diaper bags, pushing $1,000 strollers, all while dressed in the latest athleisure trend. When my son was three, I sat observing one of these moms who was playing with her toddler. As her son waddled up for a snack, she pulled out hand sanitizer and antibacterial wipes, wiped the spout of his sippy cup, deftly swiped his face, and squirted sanitizer on his hands. Then she carefully kissed him on the forehead before handing him food and water. Meanwhile, my son ran over to me and clapped my face between his visibly dirty hands. I laughed and proceeded to kiss him on the mouth, thinking all the while, *I hope this gives you oral herpes, son. You'll thank me later.*

Over the past several years, the harms of excessive hygiene for kids have been widely discussed among scientists and physicians alike. The current consensus? Let kids be exposed to microbes when they are young; it will spare them allergies and inflammation later. I'd like to add oral herpes to the list. In developing countries, almost everyone acquires oral HSV-1 during childhood. But in industrialized countries such as the United States, HSV-1 is on the decline, down from about 60 percent of the population in 1994 to just under 50 percent of the population in 2016; declines are most dramatic among adolescents and young adults.[15,37–39]

So why would I want my son exposed to oral herpes now? It's in the best interest of his penis later in life. If he develops antibodies to HSV-1 as a child, those antibodies will protect him from getting genital HSV-1 when he receives oral sex for the first time. Let's say he also happens to contract HSV-2. If he already has antibodies to HSV-1, his initial outbreak of HSV-2 will tend to be milder.

As more teens and young adults start having oral sex without these protective antibodies, genital herpes from HSV-1 is becoming more common. In the failed Herpevac vaccine trial to prevent HSV, HSV-1 infections were twice as likely to occur among young women as HSV-2.[40] And studies in Australia and the United States by Craig Roberts and Anna Wald showed that the majority of new genital herpes cases were caused by HSV-1, not HSV-2.[41,42]

While the landscape of HSV-1 is shifting, HSV-2 is beginning to disappear. Recall that one in five American adults had HSV-2 in 1994. In 2015 and 2016, the CDC estimated that figure had declined to one in eight.[39] This may be due to greater use of drugs like acyclovir to suppress viral shedding and transmission. While this is a good thing, Anna Wald worries that it may make the stigma around HSV-2 even worse. "It's interesting the prevalence has declined. The numbers from the CDC finally confirmed what I've known locally for a long time now—that [rates are] falling. In some ways it makes things worse I think, that there are fewer people who have it . . . It's not like 'oh yeah, everybody has it,' which was more likely 20 years ago. The chances of you bumping into somebody else who has it is less now."

After decades of work in the field of STIs and HIV, Wald reflected that "herpes is more stigmatized now than HIV." Now the challenge lies in reducing herpes-related stigma at the same time that the disease becomes less common. One thing that could help is an image makeover: encouraging people to view herpes as an annoyance, not a catastrophe.

Unlike untreated HIV, life-threatening complications of HSV-2 are very rare. But therein lies part of the problem. Because HSV-2 is so rarely lethal, there isn't the same community of support that exists for illnesses like cancer or HIV/AIDS. There are no colored ribbons, quilts, walkathons, or celebrity benefit galas (although without a doubt, there are plenty of celebrities with the virus). That's fair enough, but we should still do something to combat the stigma and suffering that comes along with HSV.

If I had my druthers, I'd stop calling it *genital herpes*. This evokes images of mutant genitals with open sores, and the majority of people with HSV-2 don't have symptoms. Even those who do have outbreaks look and feel fine in between. Let's call it *HSV-2 positive*, meaning the viral infection is there, but the symptoms may not be. Perhaps it would then be easier for people to come out and say, "I'm HSV-2 positive, and I'm fine. Look at my fabulous life."

Also, let's stop joking about HSV. For millions of people living with the infection, HSV-2 is serious as hell. Yet when it is mentioned in the media, it is often as the butt of a joke. Comedy films and shows such as *The Hangover, Pitch Perfect, Glee, The Colbert Report, Saturday Night Live, Knocked Up* (and nearly every Judd Apatow movie) all use the term *herpes* as fodder to generate laughs.

Would we ever joke this way about cancer or HIV? Of course not. Then why do herpes jokes continue to flourish? As blogger Leah Berkenwald aptly put it, "The jokes generally go unchecked since those who find them offensive or cruel are silenced by the fear of association with genital herpes, or the fear of being exposed as having genital herpes. Both outcomes carry the very real risks of shame, judgment, and rejection."[43] Before I became more enlightened on the subject, I joked about it too, not realizing my jokes might land on one of the millions of people who struggle with their diagnosis and must cope with it for the rest of their life.

Still, I loathe robbing comedians of good material, and these days we could all use a good laugh. But there are plenty of other funny STIs. Why don't we joke about chlamydia or crabs? They sound great rolling off the tongue and are easy to cure. Imagine receiving a card with the greeting "Congratulations! You Cured Your Chlamydia" or "In Sympathy for Your Crabs." Trust me, it would be hilarious.

2

Bushwhacked

Untangling Pubic Hair and STIs

Where Have All the Flowers Gone?

While economic forces typically drive deforestation of rain forests, the widespread deforestation of pubic hair is driven by a modern sensibility that hair-down-there is somehow unattractive or unhygienic. I can't tell exactly when this shift occurred. At some point during the early 2000s, I noticed that patients' pubic hair simply began disappearing. Not just on the sides of people's groins, where errant hair might poke out of swimwear—no, it was all gone.

First, women began apologizing if they were overdue for a wax or shave, as if the presence of hair would somehow offend me. Then men joined the ranks. Regardless of race, gender, or sexual orientation, droves of people had decided that pubic hair was the enemy. I felt quite the opposite. I missed the lush beauty of pubic hair, and I wanted it to come back.

Personally, my aversion to pain and a million competing priorities had always kept me away from bikini waxing. Occasionally through the years, I would be tempted, because come on, everyone was doing it. Then I'd read some terrifying personal account, like Christopher Hitchens's 2007 *Vanity Fair* piece where he waxed his "sack, back, and crack" at the NYC salon that originated the Brazilian.[1] He described the experience as akin to "being tortured for information that you do not possess" along with a "sandpaper handjob." This was a sufficient deterrent for me. I went along in life doing the minimum depilation necessary to avoid public embarrassment. I was happy this way for decades.

As I delved into the science of pubic hair and STIs, I began to feel uneasy. I felt that the bikini-waxing-industrial-complex was manipulating us all into doing something that could be hazardous to our health. All aesthetic self-improvement activities come with some medical risk, but I didn't know where waxing fell on the spectrum of risk between a pedicure and liposuction. Also, who was I to criticize bikini waxing when I had never really taken it off myself? What if there were some hidden benefits to pubic hair removal that I wasn't appreciating? I decided to suspend my scientific judgment on the issue until I experienced how the other (hairless) half lives.

At a training for infectious disease scientists in Washington, D.C., I met Dr. Patricia Garcez, a lovely thirtysomething Zika virus researcher from Rio de Janeiro. Somehow, our dinner conversation turned to my line of work and my interest in pubic hair. When she learned I didn't wax my bikini area, her brow furrowed. Her warm brown eyes gazed at me with concern, as if to say, "I feel sorry for your husband." When I told her I'd be traveling to Rio for the International Society for STD Research conference, she helped me devise a plan. She assured me that there would be no better place to be waxed than by the paragons of depilation themselves, the Brazilians.

After an overnight flight, I arrived in Rio on a Sunday morning and ventured over to the beach across from the hotel. Women in bikinis strolled casually along the shore. The bikini bottoms were minuscule; some were strips of fabric a few inches across, held together by strings. Everyone had more than half of their bottoms exposed. I was delighted to see that the size of the bikini had no relationship to the fitness of the wearer. Women of all shapes, sizes, and ages were scantily clad. There was not a stray pubic hair in sight.

My colleague's go-to waxing salon was in Laranjeiras, one of the neighborhoods adjacent to the huge statue of Christ the Redeemer, looming over Rio with widespread arms. It would be comforting to wax it off with Jesus nearby, but Laranjeiras was quite far from my hotel. Luckily, there was another outpost of the same chain, Pello Menos (Less Hair), just over a mile away from where I was staying. As a testament to the importance of depilation to the city's cultural fabric, this single chain boasts forty-three outposts in Rio alone, more than double the number of McDonald's.

At the conference, I had a brief window of opportunity between lunch and the afternoon keynote lecture, so I called an Uber and headed off for the salon. My driver spoke several languages, so we managed a passable conversation in Spanish, English, and broken Portuguese. He seemed unfazed by my choice of destination.

We arrived to the correct street, but despite having the address, neither of us could find the salon. We approached the area where it was supposed to be, and all we noticed was the glass storefront of a bank and ATM. The driver rolled down his window and yelled to a stranger standing on the corner. *"Eh! Onde fica Pello Menos? Depilação?"* (Where is Pello Menos? Waxing?) I sank lower into the back seat.

The stranger gestured with his chin toward the bank, and then I saw it, a narrow concrete building with no windows on the ground floor. There was a purple door half covered by a poster of a single hairless leg. The door was open, but the entrance was dark. Thanking my driver, I entered cautiously into a dark hallway with dim light coming from two bulbs screwed into the ceiling. A nondescript sign at the end of the hall indicated that the waxing studio was on the top floor.

I stepped off the elevator into a reception area filled with natural light. The young woman at the desk greeted me warmly. I spoke the three phrases that I had rehearsed silently in the Uber on the way here: *Boa tarde. Perdão, eu não falo portugues. Depilação na virilha por favor.* (Good afternoon. Pardon me, I don't speak Portuguese. Bikini wax, please.)

There were six different bikini-waxing options. Luckily, Dr. Garcez had prepared me. She'd told me many Cariocas (Rio natives) get the *virilha cavada,* loosely translating to "deep plunge of the groin." The receptionist nodded in understanding. She directed me to a menu with pictures to help me order, like a Chinese restaurant trying to be welcoming to non-Asian clientele.

On the menu I saw that you could wax every single part of your body that bears hair in various combinations or à la carte (including just your anus, in case you were curious). For those trying to wax on a budget or spot-wax a problem area, you can do just half of certain parts, such as the lower half of your bottom. She pointed to a picture corresponding to my request. The *virilha cavada* would remove almost everything from the pubis to the tailbone, leaving a tiny strip of hair in the center, a pubic runway of sorts. For a split second, I thought about backing out, but instead, I nodded.

She handed me paper underwear and led me to the bathroom to change, then escorted me to a spacious room with exposed fluorescent lighting and a concrete floor. On both sides of the room there were rows of narrow stalls, each with a single white plastic curtain. The stalls and bare concrete walls gave off a prison-shower-room vibe. I looked around and saw at least six closed curtains. English and Portuguese pop songs played quietly in the background. I strained my ears and listened. No one was yelping in pain.

Bianca, my waxer, entered my stall wearing a white uniform that resembled surgical scrubs. Her makeup was flawless, her eyebrows waxed into perfect arches. In her hand, she carried more than a quart of wax in a plastic bucket. Smiling, she placed a surgical mask over her nose and mouth and dabbed the golden wax on my inner forearm to test it, asking, "*Temperatura?*"

I nodded—the temperature was fine. I raised my skirt, and she proceeded to move the paper underwear to the side. Then wielding a long, thin spatula in her hand, she carefully applied the wax. It felt like she was frosting a cake on my groin.

Bianca then began to grab at the wax, deftly peeling it away and my hair along with it. I was surprised that it was over so quickly and how little pain was involved. (I suspect I have less hair than Christopher Hitchens.) I looked down. It looked orderly, like the groin of someone who is in control of her life, or at least her pubic hair. Perhaps I could get used to this. I left the salon feeling a sense of accomplishment. I briskly walked back to my hotel to make it to the afternoon session of the conference. Occasionally, a warm breeze wafted under my skirt and reminded me of my recent removal.

Later that evening, my pubis sang a different tune. Angry red bumps had cropped up all over my groin. A few follicles had bled where the hair had been ripped out by its roots. For years, I had talked to patients about microscopic tears and trauma to the skin from shaving and other hair removal activities. Now I really knew what it meant. Things would settle down in a few days, and it wasn't painful, but it was not a pretty sight.

Back at home, my husband appraised my new look, raised his eyebrows, and nodded, saying, "Nice." I quipped back, "Don't get used to this." As I waited for hair regrowth, I felt more authentic as I returned to writing about pubic hair removal and its relationship to STIs. I had a different perspective now; I could see the appeal in wanting to tidy up down there, whether for one's partner(s) or one's self. But after my own experience, I feared that the trauma that people cause to their skin from depilation could be a setup for STIs, and ultimately, I'm not sure that it's worth it.

I know I'm unlikely to change the hearts and minds of those who love to groom their pubic hair. I am simply here to defend it, to warn of the dangers of cutting it down. Call me the Lorax of pubic hair. I speak for the trees, for the trees have no tongues.

We're the Replacements

The practice of pubic grooming is not unique to modern times. We've been hacking away at the bush for thousands of years. The ancient Egyptians and Indians removed body and pubic hair using shaving and abrasion techniques.[2] For observant Muslims, shaving of pubic hair for both men and women was one of the *sunan al-fitrah* (way a human being should be) prescribed by the prophet Muhammad.[3]

From 4000–3000 B.C., the ancient Turks may have been the first to create chemical preparations to dissolve hair, similar to modern-day Nair. This mixture, known as *rhusma,* was made from arsenic trisulfide, quicklime (to render it alkaline and break down the hair), and starch.

Jill Burke from the University of Edinburgh found similar versions of this recipe in texts from the medieval times to the Renaissance, including specific use instructions:

Boil together a solution of one pint of arsenic and eighth of a pint of quicklime. Go to a baths or a hot room and smear medicine over the area to be depilated. When the skin feels hot, wash quickly with hot water so the flesh doesn't come off.[4]

Flesh coming off? I certainly hope pubic depilation was not as widespread back then as it is today, as the risks really seem to have outweighed the rewards.

At least some people in this period must have had pubic hair, because it was during this time that pubic hair wigs—or merkins—first made their debut. *The Oxford Companion to the Body* describes the first appearance of the merkin in 1450.[5] Sex workers of the time suffering with lice or syphilis could don a merkin to hide a shaved or diseased pubis and get back to business.

These days, merkins come in handy for actresses who need them to look authentic in period movies or in case of a color mismatch between the actress's head hair and pubic hair. They're great for drag queens looking for some coverage up front. Sometimes they are simply a fashion statement, as Lady Gaga demonstrated at a 2011 music awards show, when she sported a wavy blue-green merkin over her pants, complete with highlights.

There's now even a burgeoning merkin marketplace on Etsy, with well over a hundred options for those who want a merkin with that handmade touch. For the budget conscious, there are faux-fur pieces backed with adhesive plaster for under ten dollars, and for the fancy, it stretches all the way to the crème de la crème, a merkin with real human hair (source unclear) woven onto lace for ninety-five dollars. For those that wish to use their pubis to make a political statement, there's even one called "Make Merkins Great Again," a bushy wig with a distinct Trump-blond color and side-parted coiffure.

I think this resurgence of merkins must have resulted from the sense of desolation people felt after looking at their pubic areas post–hair removal. Hair gives texture and visual interest, so taking it away seems to have necessitated some sort of replacement. From a health perspective, other than an unlikely mishap with adhesive, the merkin trend seems pretty low risk.

Then I discovered an even stranger replacement for pubic hair, which involves encrusting one's bare pubis with crystals, a practice known as *vajazzling* or pejazzling, depending on the adjacent genital organ. I had zero temptation to try this myself, but I thought hard about whether there'd be any harm in the practice.

Unless one is allergic to the adhesive used to affix the jewels, the primary downside I could imagine would be causing abrasions to one's self or one's partner. This would be more likely if one went with real Swarovski crystals rather than plastic jewels. Most online accounts describe nuisance issues during sex. The crystals could scatter, causing a mess on the bed, or getting into both partners' hair and various orifices; one person even found one in the dishwasher. The other downside would be cost. A Brazilian wax plus vajazzling costs almost $100 at my local salon. If you need a vajacial (a vaginal facial) due to irritation/ingrown hairs from waxing, the total would be $170, and then the hair grows back in a few weeks. I'm not sure if the value is there.

I had a slight change of heart when I read writer Rachel White's account of the self-loathing and shame she felt at being diagnosed with genital warts. For the month following her diagnosis, she isolated herself from her friends. She avoided dating because she might meet a new partner and have to disclose her STI. She finally rid herself of boredom and depression by vajazzling her warts with a sparkly heart. The act of beautifying the warts made her feel empowered and reconnected with her body.[6]

I found myself wishing I could hand out sheets of adhesive crystals when I diagnosed people with warts. Why not have a little razzle-dazzle whilst you have an STI? I haven't told people to do it, because without having some scientific data, I don't like to advise someone to mess with their genitals. But if I ever get warts, I'll be the first in line at the craft store.

While the fun factor is undeniable, I suspect pubic hair replacements have not caught on widely among the general population. I hope it stays that way. If merkins or vajazzling become the new norm, I may feel forced to try decorating down there. But beyond my own selfish motives for hoping this does not become commonplace, there is something more important at stake. For some creatures, pubic hair is literally a matter of life and death.

Save the Crabs

According to the U.S. Fish and Wildlife Service, there are currently 1,456 endangered and threatened animal species, plus 29 additional candidates in line to make the list.[7] One criterion for candidacy is that the species experiences "present or threatened destruction, modification, or curtailment of its habitat or range."

For *Pthirus pubis*, the louse commonly referred to as "crabs," it is clear that habitat destruction is rampant. In the United States and Europe, 70–80 percent of adults remove their pubic hair to some degree. The frequency and extent of grooming is greater among younger adults, and universities appear to be ground zero for depilation. In one study of grooming at public universities in Indiana and Georgia, 95 percent of students had removed their pubic hair at least once in the past month. There was a broad range of removal practices, ranging from light trimming to complete slash-and-burn. Researcher Debby Herbenick from Indiana University summed it up best: "Pubic hairstyles are diverse."[8]

Hairlessness abounds regardless of age. In a study of over 7,500 U.S. adults aged eighteen to sixty-five, almost one in five qualified as an "extreme groomer," someone who removes all their pubic hair more than eleven times per year. Another one in five were "high-frequency groomers" who perform some degree of removal either daily or weekly. Yes, there were even "extreme high-frequency groomers," who remove all their pubic hair at least once a week—they made up less than 2 percent of the population.[9] My guess is few people have that kind of time.

With all this pubic deforestation in mind, I searched for the pubic louse on the U.S. Fish and Wildlife Service website in vain. Despite clear evidence of habitat destruction, *Pthirus pubis* is not considered an endangered species.

You could have fooled me. I haven't seen a good pubic louse in fifteen years. My last sighting occurred when I was a medical student rotating through the Los Angeles Free Clinic, which had a weekly afternoon clinic devoted to STIs. I saw a patient who hoped he had an STI because otherwise he worried he might be going crazy. For more than a week he'd had uncontrollable itching in his groin, yet he couldn't see any bites or a rash that would explain his symptoms. He'd bathed in hot water, slathered hydrocortisone all over his skin. Yet he just kept itching. It was getting a bit embarrassing to scratch himself so frequently, particularly in public.

To begin sleuthing, I lowered my head to get a bird's-eye view of his genitals. My eyes focused first on his penis, which looked fine. Then by relaxing my gaze, his penis faded into the background and his pubic hair came into focus. That was when I saw them. Dozens of tiny lice, clinging to his hair just as happy as you please. With his permission, I pulled out one of the hairs, laid it on a glass slide, and made my way to the laboratory.

Under the microscope, I could see the louse's squat, segmented body and claws, its latest blood meal visible through its translucent skeleton. It was a triumphant moment. When I told the patient his diagnosis, he was relieved that he wasn't crazy, and he thought I was a genius for figuring it out. That he had an STI barely registered.

I wasn't sure if my drought of crab sightings was unique, so I took an informal poll of the nurse practitioners at my STI clinic. Chuck, our most seasoned practitioner, hadn't seen them in seven years. Liz, one of our youngest clinicians, had never seen them at our clinic. Then Tae-Wol piped in that he'd seen several cases in the past few months. A recent arrival to the world of STIs, he'd previously worked in homeless and addiction medicine. I thought that was the connection; perhaps his patients still had their pubic hair. No, he assured me. The patients with pubic lice just seemed to gravitate toward him.

He also brought up an important observation. One of the patients he had seen was a pubic groomer, so there was scant hair in the man's groin. Tae-Wol had witnessed lice furiously scrambling, trying to cling to this patient's stubbly pubic growth. Then he had observed crabs in the hair over the patient's stomach and chest.

I was skeptical. Body lice, head lice, and pubic lice are different species, and each has its preferred habitat. "Those weren't body lice?" I asked.

Tae-Wol wagged his finger at me, shaking his head. "I know my lice," he insisted. These crabs had been displaced from their genital home; they had migrated along a trail of hair from the pubis to the abdomen. Now they were refugees in a strange land.

A quick search of the medical literature turned up more

evidence of Tae-Wol's observation. In 2017, there was a reported outbreak of pubic lice among eight women in a Japanese nursing home. Six had pubic lice in their scalp hair only, while the other two women had pubic lice in their pubic hair only.[10] An infested hairbrush shared between patients was partly to blame. How lice got from someone's pubic hair to the hairbrush or from someone's pubic hair to another's scalp hair remains a mystery. I don't like to imagine the potential scenarios.

It remains to be seen whether pubic lice will successfully adapt to scant pubic hair or the finer, softer hair of the body and head. For decades, ecologists have demonstrated that species that can't cope with reduced area and increased isolation from habitat fragmentation eventually become extinct.[11] And the warning signs are there. At STI clinics in the UK, Janet Wilson and Nicola Armstrong described a 60 percent decline in the proportion of patients with pubic lice from 1997 to 2003.[12] Then Shamik Dholakia from Milton Keynes General Hospital in the UK observed the incidence of pubic lice fell more than 90 percent from 2003 to 2013 (from 1.8 percent to 0.07 percent of patient visits), while the incidence of pubic grooming increased from 33 percent to 87 percent.[13]

Coincidence? I think not.

After reading Wilson and Armstrong's publication, "Did the 'Brazilian' Kill the Pubic Louse?" the Natural History Museum Rotterdam's curator Kees Moeliker realized that there were no *Pthirus pubis* specimens in its collection of over three hundred thousand different species. He put out an appeal in October 2007 through the Associated Press, asking for a generous soul to donate a pubic louse for its collection.[14] By 2013, Moeliker still had no takers. Comedian Jessica Williams learned of his quest and accompanied him around New York to ask strangers for pubic lice in her investigative piece for *The Daily Show*, "Beasts of the

Southern Wild." In the piece, she gathered other friends of the pubic louse together to highlight its plight, including director John Waters, singer Aimee Mann, and my colleague Jonathan Zenilman.

Zenilman is a professor in infectious diseases at Johns Hopkins University in Baltimore, a city that is one of the few remaining pubic lice strongholds in the United States. He still sees pubic lice regularly, every few months or so. He agrees they are not as abundant as they used to be, but feels they are not anywhere near extinction—at least not in Baltimore. I called him up to share my concern that an entire generation of millennial nurses and doctors might never know the joy of diagnosing crabs. Did he think we should mobilize our colleagues to start a rescue campaign?

Not to worry, he reassured me. "Our crowd [the STI researchers] is always trying to put itself out of business by eliminating STIs. Between climate change [creating a favorable climate for lice] and people's sexual practices, these sexually transmitted creatures are going to be with us for a very long time."

He felt global travel would also help. Pubic lice were all over the earth, in many locales where people still sported a healthy amount of pubic hair. In the book *Infections of Leisure*, Zenilman noted that travel afforded one with a sense of adventure, periods of loneliness, and exploration away from one's home environment— the perfect setup for sex with someone new.[15] Global travel continued to rise through 2018 and 2019, and hopefully some good travel sex right along with it.[16]

Although current grooming trends may continue among adults in the United States and Europe, human hair grows back quickly (unlike Brazilian rain forests). Extreme high-frequency grooming is still rare. There is likely enough hair on enough people to keep the crab louse going for a while.

Like rising and falling hemlines, there's even a chance that pubic hair will come back into fashion. In June 2017, model Amber Rose created buzz when she posted a nude photo of herself with pubic hair on Instagram, captioned "#bringbackthebush." The next month, Mark Hay reported in *Vice* that the bush was reappearing in porn. Even if a full bush doesn't come back, *Pthirus pubis* has been with us humans for at least three million years.[17] I'm sure it has weathered worse storms than this. If the situation becomes truly dire, I found a T-shirt for order online that reads, SAVE THE CRAB LOUSE. STOP PUBIC DEFORESTATION.

If pubic lice actually become extinct, there's still hope of glimpsing them for yourself. Kees Moeliker wrote me that he finally received several *Pthirus pubis* specimens after his stint on *The Daily Show*. They are now safely preserved in his museum in Rotterdam, should anyone need to pay their last respects.

While I don't think we truly need to fear the extinction of the pubic louse, I wonder if we humans are experiencing other harms from widespread pubic deforestation. Perhaps pubic hair actually serves a purpose other than as a habitat for lice. What if it is protecting us in some way that we don't yet appreciate? If so, then what are the unintended consequences of cutting it down?

The Godfather

As I read study after study about pubic hair, I noticed the same academic institution and senior author behind each of them—urologist Benjamin Breyer at the University of California–San Francisco (UCSF). In the past few years, he and his team have cornered the market on the science of pubic grooming—studying who does it and why, and the potential complications such as injuries and STIs. Then I realized that he and I both worked at the same hospital, yet we had never met.

Because Breyer's team was dominant in this arena for several years, I had imagined him to be a sort of Don Corleone–type, muscling out competing scientists, sitting behind a mahogany desk and pulling at the strings of his pubic hair research empire. He is nothing of the sort. He is both younger and more jovial than the don, and regarding other scientists who want to collaborate on pubic grooming research, his response is, "The more the merrier. I always believe that."

Breyer's foray into the world of pubic hair was not intentional; it's a most unusual path for a urologic surgeon. His prior research focused on topics familiar to other urologists but more obscure to the rest of us, such as prostate cancer survivorship and urethroplasty, a surgery to reconstruct the penile urethra after trauma.

It was his expertise in urologic trauma that prompted him to look at the National Electronic Injury Surveillance System (NEISS) around the issue of genital injuries. He knew he'd find the usual suspects: penetrating trauma from stab or gunshot wounds, blunt trauma from bicycle seats. Then he was quite surprised to find that 3 percent of adult emergency room visits for genital injuries were related to pubic grooming.[18]

Meanwhile, Breyer was also hearing the buzz about decreases in pubic lice from grooming and possible links between grooming and viral STIs from case reports in Europe. It occurred to him that pubic grooming might be doing more than just causing injury. Because grooming was often the preparatory act for a sexual encounter, it might also be involved with STI transmission and acquisition.

That's when he decided to go all in, designing a large national study of 7,580 adults that led to our awareness of "extreme groomers" (as discussed earlier) and explored the association between pubic grooming and STIs.[9]

During the study, Breyer's fellow Charles Osterberg found

that groomers were twice as likely to report a past STI compared to non-groomers, even after factoring in age and number of sexual partners. The differences were greatest for STIs that can affect the skin, such as herpes, HPV, and syphilis. However, there were significant differences even for STIs like chlamydia and HIV, which do not have an obvious relationship to grooming. That is a bit of a head-scratcher. There are many possible explanations—groomers might engage in other risky behaviors, might be more sexually active overall, or are simply more likely to recall and disclose a prior STI, just to name a few—and ultimately, it's unclear whether an STI diagnosis happens first and drives people to groom or grooming contributes to a higher risk of STIs.

When this analysis on grooming and STIs was published in 2017, Breyer didn't anticipate he'd wind up in the midst of a pubic hair media frenzy. *The New York Times* and *The Late Show with Stephen Colbert* were knocking on his door; his research was even discussed on *Saturday Night Live*. More than two hundred news outlets picked up the story, and he was barraged by press requests. On top of that, members of the public reached out to him to share their bizarre tales of pubic grooming. At some point, the attention became overwhelming, and he had to start turning journalists away. He was trying to run a urology program and a residency-training program. Pubic grooming was supposed to be a side gig, not his entire academic identity.

Breyer continues to have a scientific interest in pubic grooming and STIs, but he acknowledges that it is not the easiest way to build academic clout in the field of urology. "The work definitely lends itself to jokes. I feel like I have to defend it, and at the same time, I acknowledge how almost preposterous it is."

Breyer's work hasn't been universally accepted at scientific urology meetings. Perhaps it's because his research represents a return to the past that most urologists care to forget. Urologists were widely known as "clap doctors" (for treating gonorrhea and other STIs) at the turn of the twentieth century, but they worked for over one hundred years to change this image and position the field as a serious surgical specialty. This meant disassociating themselves from STIs and related research. In transcripts from the American Urological Association dating back to 1912, urologists lamented that they were pigeonholed into taking care of patients with STIs—one even complained that his research was largely devoted to pubic lice.[19] Another refused to have anything to do with STIs: "I will not come out and take care of the venereal cases and put urology on the basis of venereology."

While the field of urology may not wholeheartedly embrace his work, it's definitely a topic the masses are interested in. Breyer was never seeking the limelight, but he learned that "if you want to get something in the lay press, write about pubic hair, because, boy, they are just ravenous for it." Another unintended consequence of Breyer's moment of fame is that he received offers of collaboration from scientists all over the globe who are interested in his work. He now has a whole cadre of students, residents, and fellows who are eager to help him publish scientific manuscripts and have no qualms about using pubic hair as a springboard to academic success.

One of his pubic hair progeny is urologist Tom Gaither, who decided he wanted to dig a little deeper into the topic of depilation and STIs. He took Breyer's 2017 survey further by interviewing patients for their grooming and sexual practices, then actually testing them for STIs (rather than relying on recall of past infections). In 2020, Gaither reported that grooming in and of itself didn't increase risk, but that high-frequency groomers (removing all hair six to ten times per year) or extreme groomers (removing all hair eleven or more times per year) were about twice as likely to have genital STIs than those who never took it all off. [20]

It turns out that how much and how often you groom are not the only factors influencing your genital geography. Method, care, and technique during depilation also play important roles. And in some unfortunate cases, a hasty grooming session can leave you with much more than you bargained for.

Sideways

Josh was a little pressed for time; the dinner shift at the restaurant would be starting soon. His chef's whites and houndstooth pants were spotless as he sat down across from me. Another clinician had taken care of him more than a month ago for a few HPV-related warts above his penis. It had been a quick visit; the nurse had frozen three small warts with liquid nitrogen. He instructed Josh to come back if his warts hadn't gone away after a few weeks.

His warts had behaved according to plan; they had shriveled up and scabbed over. With things under control, Josh decided to tidy up his pubic hair in preparation for a date. He'd taken a razor and quickly shaved all the hair above his penis. Unfortunately, the date was a dud. Then things really took a turn for the worse. By way of explanation, Josh lowered his pants while I aimed a spotlight toward his groin.

Someone had laid flesh-colored shag carpeting above his penis. At least that's what it looked like. Each tuft of the shag carpet was a small fingerlike wart, covering the area that was frozen before and spreading across his pubis. There were at least two dozen warts now—I thought it rude to actually count while he was standing there, looking miserable.

Although Josh's warts were nearly gone when he had shaved, HPV had been lingering, either in his skin at the previous site of the warts or in the surrounding hair follicles. While shaving, HPV had contaminated his razor, and then the razor had created dozens of microscopic tears in his skin. Like tilling fertile soil and spreading seeds, his razor had unwittingly helped him create a huge new crop of warts. It was pubic landscaping gone sideways. It was also beyond what we could treat in the clinic. Josh needed a specialist with an electrocautery device and a lot of patience to burn the warts off.

After sending him off with a referral, I shared the story with our nurse practitioner, Chuck. "I know, dear," he replied in his soft Southern accent. "I always tell 'em to trim or clip, not to shave, but they don't always listen."

Shaving may be a particular culprit for both HPV and molluscum contagiosum, a common viral skin infection that can also be sexually transmitted.[21–23] Regular shaving removes the protective outermost layer of skin called the *stratum corneum,* which is Latin for "horny layer" (I did not make this up). Among patients with molluscum or HPV, patients who shave tend to have a higher number of lesions than patients who don't. It makes sense; the spread of molluscum and common (nongenital) warts is known to happen after people shave their faces, and the consequences of shaving the pubis are no different. It could be the act of shaving itself, or perhaps hair is protective, guarding the pubis from friction and microscopic trauma. Shaving off that protection may be setting people up for infection with molluscum and warts.

The problem is that shaving is the cheapest and most convenient hair removal technique around. About half of American pubic groomers use a handheld razor for depilation, about one in four use an electric razor, and one in five use scissors. The remaining groomers (less than 5 percent) use wax or laser/electrolysis; those require a lot more money and advance planning.

The combination of genitals, sharp objects, and difficult visualization is often a recipe for trouble. Based on data from NEISS, the number of emergency room visits for pubic grooming injuries increased fivefold from 2002 to 2010; a third of the nearly twelve thousand injuries occurred in the last year of the analysis. According to a 2017 U.S. study, one in four pubic groomers have sustained a grooming-related injury.[18] Most commonly these were lacerations, followed by burns and rashes. Among a study of women in Texas, more than half reported a grooming-related complication. Obese women who groomed had it even worse; they were nearly twice as likely to have a pubic grooming injury as non-obese women.[24]

If injuries are so common, why do people keep grooming with such vigor? Motivations vary according to gender, but there are some common threads. Grooming studies by Debby Herbenick and colleagues in 2013 and 2015 found associations between complete hair removal and cunnilingus.[8,25] Presumably, grooming was performed in preparation for receiving oral sex from a partner, like mowing the lawn before hosting a picnic in your yard. A study by Tami Rowen at UCSF confirmed that among women who groomed in anticipation of special events, the most commonly cited reasons were preparation for sex, followed by vacation (I'm guessing swimsuits and sex), followed by a visit to their health care professional.[26] This last reason was disappointing. Most health care providers would feel terrible if a patient sustained a grooming injury while trying to avoid some perceived offense to us. Take it from me: your pubic hair is the least of our worries.

Among men, Tom Gaither at UCSF found similar motivations for grooming: most male groomers removed hair in preparation for sex, followed by hygiene, or simply because it was part of their routine upkeep. It may come as a surprise that grooming to "make my penis look longer" did not even make the top five.[27] But Gaither did discover that the gender of men's sex partners also played a role. Gay and bisexual men groomed more frequently, and more extensively (including the anus, scrotum, and penile shaft) than straight men, who tended to focus their grooming above the penis. With greater grooming frequency and more surface area being groomed, gay and bisexual men were more prone to grooming-related injuries than their straight counterparts.[28]

With all these injuries flying around as a result of grooming, I wondered what Benjamin Breyer might say to friends and family who ask him for pubic hair advice. "You know, I think it's a personal choice and people should do what they want. The only thing I would give guidance on is, if you're going to shave off all of your hair, I would wait some period of time (at least six hours) before you have a sexual encounter."

Breyer felt a waiting period would be particularly important before having sex with someone you weren't familiar with, because of a theoretical risk of STIs affecting broken or traumatized skin. This would require some advance planning, not a frantic shave immediately before a sex act. Beyond the waiting period, Breyer didn't see grooming itself as a huge problem and said he typically gives a green light to those inclined to groom.

He reflected, though, that a neat pubis was definitely not worth the stress of a trip to the emergency room. "If you are getting nicked and having to see the doctor a lot, maybe you should change your practices." So should people with repeated injuries change their attitude toward pubic grooming?

"Yes, definitely. Maybe start keeping your geography covered."

3

The Garden of Good and Evil

Exploring the Wonders of the Vaginal Microbiome

It Rhymes with *Vagina*

If you have a name that is easily mispronounced, then you've surely developed a way to explain it to the world outside your usual circle. This is useful for doctor's appointments, restaurant reservations, or any other circumstance where a stranger says your name aloud. My routine is simple. The first time someone trips on my name, I'll explain, "It's 'eye-nah,' with a long *I* sound." If it's a situation where I will never interact with the person again, I don't bother correcting them. When it's someone I'll interact with on multiple occasions, I'll invest the time to help them get it right.

A few years ago, my friend Melanie's parents were visiting from Nebraska. I had met them twice and they had trouble with my name on both occasions. This was the third time we were meeting, and I could tell they were still struggling with my name. I gently threw them a lifeline, "It's Ina. It rhymes with *vagina*."

They sure know it now. Bet some of their friends in Nebraska do too.

When your first name rhymes with *vagina,* other kids figure it out quickly and tease you to no end. If you are scrawny and bookish like I was, beating up on these kids is not a viable option (I did try this once and failed). Instead, I developed a thicker skin and dreamed that someday I'd be so successful that all these bullies would be sick with envy, wishing they too had a name that rhymed with part of the female genital tract.

These were the pre-Google days, so it wasn't until college that I actually looked up the origin of my name. I sat on the floor of Cody's Books in Berkeley paging through a book of baby names, where I discovered that Ina is the Irish form of Agnes, meaning "pure and virginal." Ah, my mother and her wishful thinking.

Early on in my college career, the name Ina Vagina was quickly discovered and circulated among my new friends. A few dropped the beginning and ending all together and just called me Vag. It was endearing, not painful—I owned it now. However, it was jarring to hear "Hi, Vag!" yelled across Telegraph Avenue while accompanied by my visiting parents. They shot me disapproving looks. My father muttered, "Nice friends," but I just shrugged at him, saying, "Your fault, not mine."

It's clear that a career involving vaginas was my destiny, and I'm happy with how things turned out. Now I would be pleased if people imagined a huge pair of labia writing the rest of this book. Perhaps it's inevitable that I would devote a chapter of this book to the marvels and maladies of the vagina. It's a complex world in there, so I tapped my friend Jeanne Marrazzo at the University of Alabama–Birmingham to serve as my guide. Her career path into the vagina was not preordained from birth (her first name rhymes with *weenie*). It was a conscious decision by a woman determined to put lesbians and their vaginas on the radar of the scientific world.

The Lone Wolf

Growing up as a gay woman in the 1980s, Jeanne Marrazzo didn't give much thought to lesbians and their sexual health. It was a time when gay men all over the country were dying of AIDS, so early in her medical training, Marrazzo decided to become an infectious disease specialist and join in the fight. For years, her work was "defined around the sexual identity, behaviors, and health of gay men. All of us who cared for those guys who died, for anybody who had AIDS, you knew their intimate details and everybody was on board with it. You could not ignore the sex part of AIDS. It was clear how it was happening."

As she looked into launching her own research career in STIs and HIV, she realized the same could not be said for women who had sex with women. There was no fatal illness like AIDS forcing anyone to understand what women did in bed with each other and whether those activities might lead to STIs. (There have been cases of HIV transmission between women, but it is very rare.)

No one in the scientific community had given it much thought either, because the consensus around the topic was: no penis, no risk. Marrazzo understood. "It didn't really make sense from what we knew about classic STDs that lesbians would be at risk from getting STDs from other women."

Then in 1990, she found out that a friend's ex-lover had cervical cancer, a woman who had never had sex with men. She knew that the human papillomavirus (HPV) had caused the cancer, but she didn't know whether lesbians were at risk for HPV and other STIs, and if so, did any of them even realize it?

"Here I was, a lesbian heading into a career in infectious diseases, and I felt a bit embarrassed that I didn't know anything about it," she said. "So my first interest was to start talking to women who have sex with women and ask them whether they got Pap smears. It became clear very quickly that most lesbians felt like they didn't need to really worry about it. They weren't going to their physicians for contraception, so they had no opportune moment to get a Pap smear or STD testing."

In her first study, she found that among women who had sex with women, one in five had at least one type of HPV and one in seven had an abnormal Pap test.[1] These women were also less likely to report getting Pap tests than heterosexual women, because neither they nor their doctors thought they were at risk. Marrazzo's work informed current guidelines that recommend all women be screened for cervical cancer in the same way, regardless of sexual orientation.

She would have continued down the path of HPV research, but another finding diverted her during the course of her study. In addition to the Pap and HPV tests, all the women had provided samples of their vaginal fluid that Marrazzo stained and examined under the microscope. This led her to the discovery that about one in four women who had sex with women had an overgrowth of anaerobic bacteria (not requiring oxygen to survive) in their vaginal fluid. She knew that this overgrowth could cause excess discharge and odor, a condition known as bacterial vaginosis (BV).

Whether a woman identifies as straight, lesbian, or bisexual, BV is a common disorder. According to the CDC, about one in four U.S. women has vaginal bacterial changes characteristic of BV, yet many don't have any symptoms.[2] For women who do, the discharge and odor are hard to ignore. The anaerobic bacteria of BV produce unfortunate by-products, such as *putrescine* and *cadaverine,* the same chemicals that become airborne when fish begin to spoil. The age-old jokes about fishy-smelling vaginas, albeit unkind, describe a true process that occurs in BV.

A fishy vagina is not the worst of it. In pregnancy, BV increases the risk of premature birth, low birth weight, and infections of the uterus after delivery. The overgrowth of anaerobic bacteria can move from the vagina into the uterus and fallopian tubes, causing pelvic inflammatory disease.[3] To add insult to injury, BV can increase one's risk of catching chlamydia, gonorrhea, herpes, and catching or transmitting HIV.[4] These consequences can happen to anyone with BV, fishy odor or not.

No one knows exactly how long BV has been plaguing women and their vaginas. In 1913, gynecologist Arthur Curtis noted that women whose vaginas lacked a bacterium known today as *Lactobacillus* yet had high levels of anaerobic bacteria suffered increased vaginal discharge. This phenomenon only occurred among married women, so scientists felt that it must be related to sex (I suspect there was less sex out of wedlock in 1913 than there is today). The condition did not yet have a name, and the root cause of it was unknown. In 1955, the year of Curtis's death, microbiologists Herman Gardner and Charles Dukes identified a prime suspect. They believed that BV was caused by a new species of sexually transmitted bacterium they'd found frequently among BV sufferers, known today as *Gardnerella vaginalis*.[5]

To prove their point, Gardner and Dukes embarked on two ethically questionable studies. In the first of these, they inoculated the vaginas of thirteen healthy volunteers with pure cultures of *Gardnerella* bacteria that they had grown in the laboratory. Unexpectedly, only three of the women had *Gardnerella* grow inside their vaginas, and only one of these had the classic symptoms of BV.

Then Gardner and Dukes took the actual vaginal fluid from women with BV and put that fluid into the vaginas of fifteen healthy volunteers. Eleven (73 percent) of those women developed odor and increased discharge of BV, which didn't resolve spontaneously after four months. What was it about sharing vaginal fluid that caused BV while *Gardnerella* alone did not?

Gardner and Dukes never figured it out, and their experiments could likely not be repeated under today's human research standards. Here's where Marrazzo's study nicely fit the bill. By studying women who had sex with women, Marrazzo had set up a natural experiment—women sharing vaginal fluid during sex—that was undoubtedly more fun than being inoculated by men in lab coats.

Among the fifty-eight female couples in Marrazzo's study, she discovered something that took her by surprise. Nearly all (95 percent) of the couples' vaginal bacteria seemed to match up with one another.[6] If one partner had BV, the other woman in the couple usually had it too. The converse was also true: if one partner's vagina was healthy, chances were the other partner's was as well. It seemed that BV was being sexually transmitted between women,[7] and couples with BV almost always had *Gardnerella*—but Marrazzo suspected there were many other shared bacteria she hadn't yet identified that could be culpable as well.

Up to this point, Marrazzo's BV research had been done on a shoestring budget. To figure out what was really going on inside of women's vaginas, she would need more research participants and a lot more money. In 2001, she decided to apply for grant funding from the National Institutes of Health (NIH), but she worried they would reject her proposal. It was a time before science had awoken to the idea that lesbians might have unique sexual health needs. The only other scientists she knew who were interested in this topic were halfway across the world in Australia. Her boss at the University of Washington and some of her colleagues did not think it was worthy or interesting to study vaginal health, especially for women who had sex with other women.

"Part of the reason people didn't think it was important was because they just don't really think that women have sex, particularly penetrative sex," she said. "Their impoverished phallocentric imaginations precluded them from thinking about it. I had to graphically describe how women would actually transmit vaginal fluid to each other."

In her grant to the NIH, Marrazzo explained in explicit detail how women penetrated each other with fingers and sex toys, how that mixed the bacterial compositions of their vaginas, and how that might leave some of them vulnerable to BV. Because she knew that hetero women also suffered from BV, she included them in her proposal for good measure. (She left out language about "impoverished phallocentric imaginations.")

Marrazzo also proposed something other scientists had not yet been able to do. Previously, BV researchers used culture techniques to grow out different species of bacteria from the vagina. This would involve taking samples of vaginal fluid and adding them to a liquid broth with nutrients that would encourage bacterial growth. The bacteria that could be identified were the ones that would actually grow under these conditions. Marrazzo suspected there were many more species of vaginal bacteria yet undiscovered because they needed the vagina's unique environment to thrive: a warm, moist, oxygen-poor cave.

While she was preparing her grant, she became acquainted with David Fredricks at the University of Washington's cancer research center. He was pioneering the use of rapid genetic sequencing to describe bacterial communities in exquisite detail all over the body, popularly known today as the *microbiome*. With Fredricks's help, she knew she'd be able to create more detailed profiles of the vaginal microbiome than had ever been described. She called him up, saying, "I think you should study the vagina. It's really interesting." He agreed.

The NIH funded her grant proposal, launching her into a

successful career studying the microbiome of the vagina, nearly a decade before the human microbiome became *the* hot topic on *Dr. Oz,* in popular science books, and in every wellness guru's blog. David Fredricks became her partner in crime and the vaginal microbiome expert for the NIH's Human Microbiome Project, a $157 million, five-year effort to study the microbiome and its relation to health and disease.

Three years later, Marrazzo and Fredricks were able to identify almost four dozen different types of bacteria in the vaginal microbiome, including three new bacteria that had never been described (they named these *BV-associated bacteria,* or *BVAB, 1, 2, and 3*). They published their discoveries in the prestigious *New England Journal of Medicine* in 2005.[8] Marrazzo refrained from gloating to those who said the vagina was unworthy of her attention.

As Marrazzo and Fredricks discovered during their years together, the vagina is a wondrous, complex ecosystem. At its best, its microbiome adapts to shifts in hormones and the monthly presence of menstrual blood. It accommodates intruders like sex toys, lubricants, penises, fingers, or semen, and the bacteria that might come along with them—its microbiome handles these insults and quickly bounces back to its optimal healthy state. But every ecosystem has its vulnerabilities, and the vagina is no exception. And if a vagina's ecosystem goes to the dark side, bringing it back can be no mean feat.

There Goes the Neighborhood

Whether you happen to be a fan of the actress Awkwafina (*Crazy Rich Asians*) or not, she's a genius at capturing how one might feel when their vagina is in peak health. In one of the opening scenes of her video "My Vag," you see her dressed in a white coat, donning a surgical head mirror, poised between a pair of legs draped with a sheet. Throughout the video, she pulls various unlikely items out from between a pair of legs: a Big Gulp cup, yellow CAUTION tape, and a wet mop, all while rapping about the virtues of her vagina.[9]

> My vag, like tasting heaven
> Yo vag manages a 7-Eleven
> My vag, it's ornamental
> Yo vag is a five-hour PBS special

(I agree with Awkwafina's assessment except on one point: she needs to watch the six-hour PBS/BBC version of *Pride and Prejudice*. When she does, she'll realize the tremendous honor of having her vagina associated with public broadcasting.)

When things are going well in the vagina, millions of good bacteria, certain species of *Lactobacillus,* happily reside on the vaginal walls and within vaginal fluid. There they metabolize the sugar molecule glycogen inside the vaginal tissues, producing lactic acid as a result. In a happy vagina, the lactic acid from *Lactobacillus* creates a pH between 3.5 and 4.5.[3] This means vaginal fluid should taste slightly tangy, but not sour: think yogurt or merlot, not lemon juice.

Unlike the gut, where a wide diversity of bacteria is a plus, the optimal vaginal microbiome has almost all *Lactobacillus* and little else. There are dozens of other bacterial species that can potentially grow in the vagina, but ideally, the acidity of the environment keeps those bacteria under control. The acidic conditions also help protect against bacteria or viruses introduced by fingers and penises, including STIs and HIV.

Even though people are highly diverse in age, race/ethnicity, and sexual behavior, their vaginal microbiomes can typically be classified into one of five major categories, called *vaginal community states I* through *V.* These include optimal and less optimal environments for sexual and reproductive health. In the optimal community states (I, II, III, and V), *Lactobacillus* dominates, the pH is nicely acidic, and all is right with the world.

Community states on the other end of the spectrum are different. In community state IV, *Lactobacillus* may be present, but they are less plentiful, allowing growth of diverse groups of bacteria who also thrive in the vagina's anaerobic environment. *Gardnerella vaginalis* is the best known among these, but there are dozens of others with names reminiscent of Cinderella's stepsisters: *Prevotella, Sneathia, Eggerthella.*

In less optimal community states, the vaginal pH is higher than the ideal range of 3.5–4.5 (closer to 5.5). These states would be present when women have the discharge and odor of BV. However, less optimal states aren't always associated with disease. Researcher Jacques Ravel found these community states in over a third of healthy Black and Hispanic women and almost 20 percent of Asian women, compared to less than 10 percent of white women.[10] Notably, his team didn't actually ask women about symptoms, and they couldn't tell whether the differences in community state were related to genetics, personal habits, or sexual practices. While *Lactobacillus*-poor community states may be "normal" for some women, these women are more vulnerable to developing discharge and odor, plus other negative outcomes related to pregnancy and acquisition of STIs/HIV.

Just like in any neighborhood, the vaginal bacterial community state can shift and change.[11] Some vaginas stay in one state for weeks, while others can change dramatically over the course of a day.[12,13] The monthly presence of menstrual blood (pH 7.4)

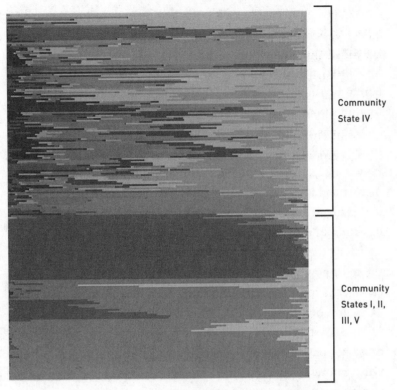

Community State IV

Community States I, II, III, V

Species of bacteria in the vaginal fluid of 220 women. Each woman is represented by a thin horizontal line, and more shades of gray represent greater diversity of bacterial species. Women with BV (microbiome often in community state IV) have more diverse bacteria compared to women without BV (microbiome often in community states I, II, III, and IV). Modified from S. Srinivasan et al. *PLOS ONE*, 2012.[14]

causes everyone's vaginal microbiome to take a hit. *Lactobacillus* concentrations go down, and *Gardnerella* concentrations go up. Vaginal pH also takes a hit when semen enters the picture (pH 7.2–7.8). Whether a vagina can bounce back to its optimal state and how quickly that happens will be different for every person.

We all know people who are stable as rocks: they lose their jobs, break up with their partners, and crash their cars, yet they remain steady and calm. (I am not one of these people, but I know they exist.) On the other hand, there are people who are barely holding it together, where you know any tiny stressor might just push them over the edge.

Vaginas are like this too. A resilient vagina can take multiple insults, and its microbiome quickly recovers to its usual state. However, for a fragile vagina, even a small insult like semen or menstrual blood can push it into a state of chaos. (Note that vaginal resilience does not appear to be related to emotional stability. You can be a hot mess and have a vagina of steel.)

Many forces can affect the resilience of the vagina, and one of the most powerful of these is estrogen. *Lactobacillus* thrive when estrogen is most plentiful, typically during the decades after puberty and before menopause. When perimenopause strikes, the roller coaster of waxing and waning estrogen levels doesn't just cause hot flashes and changes in bleeding. This hormonal instability can result in declines in *Lactobacillus*, placing a previously stable vagina in a more vulnerable state.

This isn't just a women's issue. When transgender men (born female, transition to male) begin testosterone therapy, estrogen levels in the vagina plummet, and the vaginal microbiome transitions toward a menopausal state. So just when transgender men might be enjoying the masculinizing effects of testosterone, they may face more vaginal troubles than they did before they began their hormonal transition.

Once menopause has fully set in, the vaginal community changes yet again. Without estrogen, the vagina's surface thins out and becomes drier. Its tissues produce less glycogen, which means less food for *Lactobacillus* species. And it's not just *Lactobacillus* that moves out of the neighborhood—the BV-associated bacteria often do too. A vagina once crowded with bacterial inhabitants becomes more deserted after menopause.

While the effects of menstruation, menopause, and hormones on the vagina are often outside our control, we do have power over what we choose to put inside them. The only things that have any business inside the vagina are for pleasure or menstruation: fingers, sex toys, tampons/menstrual cups, or penises (one at a time is easiest). Here's a good rule of thumb: if you wouldn't put it in your mouth, then don't put it in your vagina. (Okay, no one actually puts clean tampons in their mouth, but you could.) Unfortunately, the converse isn't true. Keep that pizza out of your vagina, as it will wreak havoc on your microbiome.

These rules to live by seem simple, yet millions of us abuse our vaginas each year by exposing them to douches and other "feminine hygiene" products. This obsession with cleaning and perfuming our vaginas is certainly nothing new. Extreme vaginal hygiene has been a long-standing American pastime—one with surprising and sometimes disastrous consequences.

That Not-So-Fresh Feeling

In 1975, at the Illinois Institute of Technology in Chicago, gynecologist Louis Keith and a team of chemists were attempting an ambitious task. They were trying to describe exactly how a normal vagina was supposed to smell.

Keith's team recruited ten volunteers to insert tampon-shaped perforated Teflon devices into their vaginas to provide samples of their vaginal fluid. Keith's team took these devices and placed them in glass tubes, which they sealed to capture the vaginal vapors. Using a process called *gas chromatography*, his team separated and analyzed each of the individual odors emanating from the vaginal

vapor. In all, his team found 2,099 unique smells or "odorous ef-
fluents" in women's vaginal fluid. They grouped these into different categories, and then graded each of them on a scale: P to PPP
for pleasant smells, to N for neutral, and X to XXX for noxious
smells.[15]

Here is a small sample of the odors that they found:

Pleasant: Fruity, floral, spicy, sweet

Neutral: Yeast, alcohol, bran

Unpleasant: Burnt earth, amine (fishy), cheesy, medici-
nal, fatty, bitter, sweaty, and (a bit of a head-scratcher)
leafy plants

Not only did each woman's vagina have hundreds of different odors, but the ratio of pleasant to unpleasant odors changed
throughout the month. Vaginal odors were most pleasant at the
middle of the menstrual cycle near ovulation and least pleasant
immediately before and after the woman's period. Keith's team
considered trying to define all the chemical compounds responsible for vaginal odor, but they realized that they were in over
their heads. Their conclusion? It was impossible to define exactly
how a "normal" vagina ought to smell. Keith wrote, "The olfac-
tory signature of an individual is complex, highly individual, and
composed of many 'mini odors.'"

Keith's team didn't realize it at the time, but an individual's
unique scent and fluctuations in that scent reflect the vaginal
microbiome community state at different times throughout the
month. These natural rhythms are interrupted during BV, when
a woman may have a constant fishy odor from the by-products of
bad anaerobic bacteria. But bottom line, a healthy vagina should

always smell like *something* (not fishy like BV), even though that something is hard to describe in words.

For over a hundred years, the feminine hygiene industry has managed to convince women that they need to mask the smell of their vaginas. The earliest and most pervasive commercial products were douches, but today, there are countless ways that you can wash and perfume the vagina: liquid washes, wipes, powders, sprays. There are even products for those vaginas that get cranky in the evenings: "Perfect for a touch up, or before getting into bed, our Night-time Cleansing Cloths give you 5 fresh benefits in 1 luxe cloth." Between washing, douching, wiping, and spraying day and night, you'd need an extra hour per day for all this rigmarole.

The practice of douching has been around for centuries, but primarily as a means of preventing pregnancy or curing infection, not simply for personal hygiene. In *The Medieval Vagina*, Karen Harris and Lori Caskey-Sigety describe douching recipes written on papyrus from 1500 B.C., which recommended the use of garlic and wine as a douching solution. The list of other ingredients in medieval douches reads like a recipe for salad dressing: olive oil, pomegranate pulp, honey, and ginger.[16]

The introduction of the douche in the United States has widely been credited to New York physician Charles Knowlton in 1832, in his reproductive health text, *Fruits of Philosophy, or the Private Companion of Young Married People*.[17] He proposed douching for contraception with a spermicidal solution of salts such as zinc, aluminum, and potassium sulfate. Other douches of the time used similar salts, plus carbolic acid, boric acid, and salicylates (used in dandruff shampoo and acne washes). For extra tingle, they threw in eucalyptus, menthol, and camphor. It should be noted that none of these solutions were actually effective as contraceptives, but in Knowlton's text, he touted the many advantages of douching: "It costs nearly nothing; it is sure; it requires no sacrifice of pleasure; it is in the hands of the female; it is to be used after, instead of before connexion, a weighty consideration in its favour."

Little did Knowlton know that New Yorkers had already beaten him to the punch at least a decade before *Fruits of Philosophy* had been published. In 2014, a two-hundred-year-old douching syringe carved out of bone was discovered in a pile of trash beneath city hall in Manhattan.[18] According to archeologists, the trash pile dated back between 1803 and 1815, right about the time that the building was being completed. The douche appeared among the remnants from a serious party: alcohol bottles, pipes, fine china, and bones from cows, fish, and turtles that were likely served for dinner. The douche, one can assume, was part of the after-party.

During Knowlton's time, hawking douches to American women was not without its challenges. In 1873, Congress passed the Comstock laws, making it illegal to use U.S. mail to advertise "erotica, contraceptives, abortifacients, or sex toys." Douche manufacturers had to get creative to market their products. In *Controlling Reproduction: An American History,* Andrea Tone describes taglines for douche ads of the time, designed to engender fear of unplanned pregnancy without mentioning the word *contraception:* "Young Wives are Often Secretly Terrified," "Can Married Women Ever Feel Safe?" or "Calendar Fear."[19]

After several studies revealed that douching was an ineffective contraceptive, douche manufacturers had to pivot again. In the 1920s and '30s, douching ads focused on making women conscious of their vaginal odor and the fallout it would have on their lives. To accomplish this, they employed a number of messages: 1) your vagina smells and you do not even realize it, 2) your malodorous vagina is at the root of your life problems, and 3) you will drive your husband to become a philanderer if you do not douche away your odor. There were many companies in the douche market at that time, including Marvel, Zonite, and Sterizol. However, the disinfectant company Lysol was particularly successful at playing to women's anxieties.

> Mrs J—— is pretty, poised and friendly. You'd think that both men and women would like to talk to her, like to have her around. But she's careless about "the one unforgiveable fault." So she's seldom invited back a second time. **Any woman** may be the victim of "embarrassing odor" *without realizing it.*
>
> "I guess I was really to blame when Stan started paying attention to other women. It wasn't that I didn't know about feminine hygiene. I had become . . . well . . . *forgetful.* Yes, I found out the hard way that 'now-and-then' care isn't enough. My doctor finally set me right. 'Never be a careless wife,' he said. He advised Lysol disinfectant for douching *always.*"

This age of vagina shaming would continue for decades. But the manufacturing of shame wasn't the worst part. Douching with a disinfectant such as Lysol was akin to throwing a bomb inside the vagina. It certainly killed any bad anaerobic bacteria, but it also killed any good *Lactobacillus* in the process. If a woman's vagina were healthy and resilient, *Lactobacillus* would recolonize and her vagina would return to normal. But if her vagina was already overrun with BV, then repeated douching *reduced* her vagina's ability to repopulate with good bacteria, perpetuating her problem instead of solving it.

For decades, there was little public discourse on the dangers of douching with harsh disinfectants. As late as 1959, there were still Lysol ads assuring women of their product's safety, promising that their "personal daintiness" would be served by douching with Lysol. By this time, Lysol offered a choice of scents: regular or pine fresh. I suppose that if one is going to decimate their microbiome, they might as well smell pine fresh afterward.

Surprisingly, in the 1960s, the women's liberation movement and the birth of the pejorative term *douchebag* didn't kill the practice. But douche manufacturers had to change their messages yet again. Now douching ads portrayed the practice as means of sexual empowerment: women were no longer douching out of shame; they were doing it to enhance their partner's sexual enjoyment.

Consider Amy, a young woman lying in her partner's lap, a sultry look in her eyes. She's clearly a liberated woman: her hair is loose and flowing, she's not wearing a bra, her nipple is visible beneath her sheer tank top. Her partner loves her just as she is, but he appreciates that she's chosen to douche.

Amy doesn't have to douche. But she knows I love apricots. LOVE APRICOTS®. One of sixteen delicious flavors-of-love from LOVE

Douching was a cockroach that simply would not die. In the 1988 National Survey of Family Growth, more than a third of fifteen- to forty-four-year-old American women reported douching the past year.[20] By now the scientific and medical communities had finally begun discussing douching's potential harms, including associations to BV and pelvic inflammatory disease. But they were up against a practice ingrained in the psyche of American women, a tradition that had been passed down from mothers to daughters for more than four generations.

Slowly but surely, the tide began to turn. By 2002, about a third of U.S. women had douched in the last year, but that fell to fewer than one in five women between 2011 and 2015. That's still eleven million women too many, but it's progress. As gynecologists and scientists recite their mantra to the next generation—the vagina is a self-cleaning oven—use of feminine hygiene products and douches will hopefully continue to decline. Today, the only folks who should continue douching their vaginas are transgender women after gender-affirming surgery. The neovagina, created from the skin of the former penis, is not a self-cleaning organ. You need to get in there and do it yourself.[21]

Douching is not necessary for women with a healthy vagina and even worse for women with BV. A vagina with BV needs medical attention, not a washout and disinfection. Today, there are antibiotics such as metronidazole that selectively kill the anaerobic bacteria of BV but leave the good *Lactobacillus* species unharmed. Once the bad actors have been cleared away, *Lactobacillus* will try to take back the neighborhood. If it succeeds, vaginal harmony will be restored. But while this is happening, the vagina is in a fragile state. It is best left alone to its own devices.

Unfortunately, once a woman has experienced symptoms of BV, she is likely to do so again. While nearly all women initially respond to treatment, about half will have recurrence after a year.[22] Women who have sex with women and racial/ethnic minority women are particularly vulnerable. Repeated bouts of BV may stretch out for years, causing unending emotional pain and frustration. The internet is rife with pleas of women desperate for help. On Reddit, BV sufferers have lamented: "BV is ruining my life." "I cannot live like this." "Please tell me if u [*sic*] know some magic trick!"

Why is it that some women's vaginas are so resistant to being cured? And when the forces of evil bacteria appear to be undefeatable, where can a vagina turn for help?

The Penis and the Great Wall of Vagina

By the time that Stella had reached Jeanne Marrazzo, she was completely fed up with her vagina and the army doctors who'd tried in vain to fix it for the past year. She'd only been having sex with one man during that entire time, a fellow soldier named Clay, who would soon be deployed overseas. He insisted that he was faithful as well.

A few months after she and Clay had started having sex, Stella noticed an increase in her vaginal discharge and an unpleasant odor, prompting her to see a doctor on base. The doctors tested for gonorrhea and chlamydia, but they doubted that either was the cause of her symptoms. After looking at a sample of her vaginal fluid under the microscope, the doctor diagnosed with her with BV and prescribed a weeklong treatment with metronidazole, the typical first-line antibiotic therapy.

Stella avoided having sex with Clay for a week in case her STI testing was positive. She was relieved when the doctor's office called to say that her other tests were negative; it was "only" BV. After she had finished her antibiotics, it seemed like her vagina was back to normal, but that feeling didn't last for long. She and Clay resumed having sex, and a little over a month later, her symptoms returned. She returned to the doctor, who prescribed antibiotics again and then again. Each time her symptoms would go away, but then a month or two later, they'd inevitably come back.

Stella decided to become her own vaginal detective. She noticed that some episodes of BV seemed to be triggered by having sex with Clay. She was on the Pill, and neither of them had other partners, so they were no longer using condoms. She wondered whether something in his semen was causing her issues.

She decided to seek more advice and made a pilgrimage to

the STI clinic in Seattle to see if she'd get some answers. Seattle's STI clinic didn't specialize in BV, but because of Marrazzo's expertise, it became a de facto refuge for troubled vaginas from all over the area. One of Marrazzo's colleagues saw Stella, and he realized that she qualified for a study they were doing of a newer antibiotic for BV called *tinidazole*, which they hoped would give her lasting relief.

Stella enrolled in the study and took the antibiotics faithfully, but two months later, she was back again with BV. Marrazzo sat down with her to troubleshoot the situation. Stella had been through treatment for BV at least six times in the last year; she was ready to try a different tactic. She asked Marrazzo whether treating Clay would help to end her recurrences.

Marrazzo explained that BV was not a classic STI like chlamydia, where treating the patient and the partner always cured both parties. In fact, scientists have attempted several clinical trials to treat male partners of women with BV: using alcohol gels, oral antibiotics, and topical antibiotics, in varying combinations.[23,24] While further trials are under way, no one has yet discovered the treatment for men that will prevent recurrent BV for women.

The inability to treat male partners of women with BV begs the question: Is BV simply a bacterial imbalance in the vagina somehow triggered by sex, or a sexually transmitted infection in its own right? Even more than one hundred years after BV was first described, no one is quite sure. When it comes to BV and sexual transmission, it is certainly guilty by association. BV usually occurs after a sexual experience, which can entail doing anything (penetration with fingers, toys, or penis) that might mix two people's genital microbiomes.[25]

And we still don't know the exact factor that pushes a person's vagina over the edge. In Stella's case, it's possible that her vagina couldn't tolerate the high pH of semen, and unprotected sex drove her vagina toward the disordered community state where BV could thrive. Marrazzo also thought Clay's penis was harboring BV-associated bacteria, and thus he would continue to seed Stella's vagina with it each time they had sex. Perhaps it was a little of both.

While the world of the penis is not nearly as complex as the vagina, penises do have microbiomes of their own. Bacterial communities can be found on the outside skin of the penis, underneath the foreskin (if it still exists), and inside of the urethra. Dennis Fortenberry from Indiana University found that the penis's microbiome changes shortly after someone has sex for the first time. For men who have sex with women, the good bacteria like *Lactobacillus* that are found in the vagina also make their way onto the surface and inside the penis.

When a man's partner has BV, the diverse BV-related bacteria from the vagina can be detected on the outside and inside of the penis as well.[26] Fortenberry wondered how these vaginal organisms actually got inside.

"The male urethra is not just a highway that you turn into and drive down," he pointed out. "Men that sit in a bathtub don't get water in their urethra. Men that swim don't get water in their urethra. There's got to be some mechanism pushing those organisms in because they don't fly, they don't have feathers, they don't have feet, they can't walk, they don't have fins, they don't swim. How do they get there? If you remember, the target is about a centimeter in width. That's the urinary meatus [the hole on the head of the penis]. That's a pretty small target. If they miss that, they miss the boat."

After much thought, he concluded that it's all about the thrust. The compression and decompression on the penis, the changes in pressure that happen during thrusting, create an ebb and flow that might drive BV-associated bacteria (and other STIs) inside the penis. The way could be paved by the pre-ejaculatory fluid that forms even before the thrusting begins. Fortenberry tried to recruit an engineer at Purdue to help him characterize the fluid dynamics of vaginal secretions as they flowed over the surface of the penis and into the male urethra. He didn't get very far. It may have been a bit too much for the engineer.

Because a man can temporarily harbor BV-associated bacteria in his penis after sexual exposure, if he has multiple female partners in a short time frame, he could potentially seed the vagina of one partner with BV-associated bacteria from another partner. Yet despite the presence of these bacteria on and inside his penis, a man will never suffer from the fishy odor and discharge that his female partners often face. (If men suffered from fishy penises, I believe we would have had the cure for BV decades ago.)

As Marrazzo counseled Stella, she didn't want to say that Clay and his microbiome were to blame. She couldn't be sure, and she was not about to cause a rift in anyone's relationship. Imagine if Stella broke up with him over it: "I'm sorry, it's not you, it's your microbiome." Instead, she proposed that she and Clay try condoms for a while to see if blocking exposure to semen and penile bacteria might help.

"They didn't want to use condoms, because people in monogamous relationships don't want to use condoms. But I tried to get her to do a trial, just to point out that it can help," Marrazzo said.

If her suffering had gone on much longer, Stella might have changed her mind about using condoms, but it turned out she didn't have to. Later that month, Clay deployed overseas, and she didn't have sex for six months. For the first time in over a year and a half, all was calm in Stella's vagina; she didn't have a recurrence of BV.

"It was complicated because she was committed to him, he was her partner, and yet probably he was provoking the change [in her vagina] or actually giving her the seeds of the BV-associated bacteria."

Clay returned from his deployment, and a few months later, Stella and her vagina were back to square one. After Stella's multiple unsuccessful treatments in the past, Marrazzo figured the usual antibiotics just weren't going to cut it. It seemed that treatment kept failing because antibiotics were only killing planktonic bacteria, those that were free-floating in Stella's vagina, much like the microscopic organisms floating in the sea. But during Stella's many bouts with BV, Marrazzo suspected that the planktonic bacteria had actually begun to stick to Stella's vaginal walls. There, they settled in, building a self-containing structure called a *biofilm*, a slimy matrix of mucus, proteins, and bacteria.[27]

Biofilms are found throughout nature in warm, moist places, but we're most familiar with them on our teeth (plaque) or as slime on the walls of our shower. While we're able to clean off these biofilms with a toothbrush or scrubber, there's no easy way to brush biofilm out of all the vagina's nooks and crannies (so please do not try that at home).

The biofilm was likely sabotaging Stella's treatment in two ways. Biofilms create a physical wall within which anaerobic bacteria can seek refuge, making it difficult for antibiotics to penetrate and kill. Then the clever bacteria enter a state of hibernation—they lie there and play dead. Bacteria within the biofilm ratchet down their DNA replication and protein synthesis, so even if antibiotics reach them, their slow metabolism make them difficult to kill.

Marrazzo realized that antibiotic treatment was clearing out Stella's planktonic bacteria, but presumed that the bacterial residents of the great wall of biofilm would just emerge from hibernation, multiply, and overrun the vagina again. *Lactobacillus* would not have a chance to repopulate. And if a penis entered that also harbored BV-associated bacteria, it would put those in the mix, throwing more fuel on the fire. Combine this with the pH disruption from semen and menstrual blood, and Stella's vagina simply couldn't cope.

The secret to a cure would lie in breaking up the great wall. To do this, Marrazzo decided to use boric acid, a throwback component of douches from the 1800s. She knew boric acid alone would not be harmful to the vagina's delicate tissues, yet strong enough to disrupt the matrix of mucus and protein created by the biofilm.

First Stella took metronidazole pills by mouth to kill off the free-floating planktonic bacteria. Then Marrazzo had her place boric acid suppositories in her vagina for three weeks in an attempt to break up the biofilm. To cap it all off, she ended with using metronidazole gel inside the vagina for four months, to keep anaerobic bacteria at bay while the good *Lactobacillus* could repopulate once again.

Finally, after almost two years of wrestling with BV, Stella was cured.

Marrazzo still uses this complicated recipe for patients with recurrent BV. It takes some commitment, but it seems to work, and it is now included in the current CDC treatment guidelines for BV. Another part of an ideal treatment would be a vaginal probiotic, where, according to Marrazzo, "You would get the bacterial burden down, wipe out the biofilm, suppress [anaerobic bacteria] for a couple of weeks, and then replenish the normal flora [with a probiotic]."

While Marrazzo's idea makes sense, vaginal probiotics available over the counter often use *Lactobacillus* from cows: great for making yogurt, but not helpful to a human vagina. But help may soon be on the way. An early (Phase IIb) study of a potent probiotic made from human *Lactobacillus* (Lactin-V) has shown promise in preventing recurrent BV.[28] Stay tuned to see if it will succeed in larger trials and make it into women's hands (and their vaginas).

What about using vaginal fluid itself? From Marrazzo's prior work, she knew that BV could be transmitted by women sharing vaginal fluid with each other. Perhaps the opposite could also be true—maybe the transfer of a healthy woman's fluids would help cure a woman with recurrent BV. A vaginal fluid transplant could be somewhere on the horizon.

Lest you think this is a crazy idea, it's already happening using something with a higher yuck factor than vaginal fluid: human feces. Cases of severe diarrhea caused by the bacterium *Clostridium difficile* can be terribly difficult to treat, resulting in hospitalization and even death. Transplanting poop from a healthy donor has been revolutionary in saving lives for people with recurrent bouts of this diarrhea. The healthy person's poop, full of good bacteria, is spread like fertilizer all over the colon of the ill patient. The healthy bacteria in the donor's poop then take hold, restoring a healthy microbiome.

Marrazzo dreams of trying out something similar with a vaginal fluid transplant. If she ever pulls it off, I will surely volunteer to donate my vaginal fluid to help a suffering stranger. There is plenty more where that came from, and it's a much lower commitment than donating a kidney.

Walk Softly and Carry a Big Stick

Although scientists and clinicians have finally aligned themselves on the dangers of douching, the feminine hygiene industry continues in their game of Whac-A-Mole. My colleague Hillary Liss at the University of Washington turned me on to a product called the *Virgin Stick,* a cigar-shaped vaginal insert made up of chalk, herbs, and who knows what else, which claims to cure numerous vaginal issues, and also happens to prevent cancer.

According to product ads, the "Virgin Stick is a natural vaginal discharge treatment that also gets rid of vaginal odor and even tightens a loose vagina. This virgin stick is made up of a combination of herbs that are popular in Indonesia to cure smelly vaginal discharge and get rid of the vaginal itching . . . It can help in the prevention of any more serious disease such as cancer (though of course you should always do everything you can to stave off disease)."

Available at multiple online retailers, the sticks come in several colors, some scented with essential oils. One manufacturer of a similar product from China promises that it will "deodorize the womb of a very nasty husband" (I suspect something may have been lost in translation).

Vaginal hygiene snake oil peddlers come in all forms, including blond, beautiful, Oscar-winning actresses. Gwyneth Paltrow's Goop site offers all manner of questionable advice on hygiene practices such as vaginal steaming, which involves sitting on a "mini throne" whilst a combination of infrared and mugwort steam cleanses your uterus.[29] Never mind that steaming your vulva and vagina will *not* cleanse your uterus. Today's Goop-ers might be wise to the fact that cleansing one's reproductive tract is unnecessary and harmful, so Goop also touted that steaming provides "an energetic release—not just a steam douche—that balances female hormone levels" (also impossible, by the way).

Then, of course, there are Goop's infamous jade eggs. Goop's experts claimed that insertion of the vaginal eggs was an ancient practice of Chinese concubines, which would balance hormones/menstruation, prevent prolapse of the uterus, and "cultivate sexual energy, clear chi pathways in the body, intensify femininity, and invigorate life force." Women were instructed to keep the egg in for hours, even as long as overnight.

Scientifically minded gynecologists became incensed, particularly Dr. Jen Gunter, author of *The Vagina Bible* and the blog *Wielding the Lasso of Truth*. Her gripe with Goop was that jade, being porous, could harbor bacteria that might increase the risk of BV or toxic shock syndrome (TSS).[30] Other gynecologists came to Goop's defense, arguing that while jade was porous, it was not absorbent like a tampon, so risk of TSS was minimal. There are no studies looking at jade eggs and the risk of TSS or BV (and hopefully no such study will ever be funded). However, placing a foreign body that seals off the vagina for hours sounds like a very bad idea to me.

The online war of words around jade eggs may have helped serve Paltrow's interests. As journalist Taffy Brodesser-Akner pointed out in *The New York Times,* critical blog posts and articles often linked to Goop's website and actually drove up traffic. Paltrow was well-poised to take advantage of the "cultural firestorms" that drove more eyes to Goop. "I can monetize those eyeballs," she told students at Harvard Business School.[31]

Goop eventually settled a lawsuit from the California Food, Drug, and Medical Device Task Force for $145,000—a measly settlement for a company that claims its worth at $250 million.[32] At the very least, consumers of the jade eggs were offered their money back, and unsubstantiated medical claims about vaginal steaming and jade eggs were removed from Goop's website. Goop currently has a full-time fact-checker, which Paltrow referred to in *The Times* as a "necessary growing pain." Today, articles about Goop's health products contain the following disclaimer:

The article is not, nor is it intended to be, a substitute for professional medical advice, diagnosis, or treatment, and should never be relied upon for specific medical advice.

No kidding.

From a pure safety standpoint, steaming or putting a foreign body in your vagina may not kill you, but realize that Goop doesn't care about your vagina any more than douche manufacturers of the 1930s did. They are using the same psychological manipulation to try to sell their products, conveying the message that you and your vagina are not good enough as is, and you need *something* to improve your sad existence. In 1959, women were told they needed to cultivate their "daintiness"—today it's couched as "femininity," but the unspoken message has stayed the same. The fact that the message is framed as empowerment doesn't change the deficit mentality behind it.

Now if one of these vaginal accoutrements happened to change your life, then more power to you. I'm not sticking a Virgin Stick anywhere or steaming my vagina, but I admit I was tempted to see if walking around with a vaginal egg would unleash my inner concubine. Then I saw Jen Gunter's 2018 study that searched for jade eggs in databases of Chinese artifacts; no evidence of the eggs could be found.[33] The Chinese concubines and jade eggs were just a marketing ploy. In the end, I could not bring myself to give Goop sixty-six dollars of my hard-earned money.

Since Goop can no longer tell you what to do, here's some specific medical advice on which you can rely: take your sixty-six dollars, put it toward a good vibrator, and enjoy a bunch of orgasms. I promise that you and your vagina will feel better.

4

Warts and All

The Omnipresence of the Human Papillomavirus

Knockout

One of my most formative life experiences came during my sophomore year of college, when I became a sexual health peer educator at UC–Berkeley. I loved counseling students one-on-one about sex and STIs, then traipsing to dorms and fraternity houses with a plastic penis to perform condom demonstrations. By my junior year, I was helping to coordinate the program, which involved selecting and training the new educators.

Every fall, we had an intensive week of training for the educators to learn the nuts and bolts of STIs, contraception, and Pap testing for cervical cancer. In the 1990s, young women usually had their first Pap smear at age eighteen. That meant hundreds of incoming freshmen would descend on the student health center to have their first pelvic exam and Pap smear. The peer educators would be tasked with seeing the students beforehand for a brief show-and-tell with the equipment and to answer any questions, and so the educator training culminated with a live demonstration of a female pelvic exam. We knew that a bird's-eye view of the exam was no substitute for living the experience, but it had to be better than nothing.

It was the day before the big day. We had hired someone to serve as a live model and arranged for an exam room in the clinic and a volunteer doctor. Many of the peer educators wanted to go on to medical school and were excited at the prospect of seeing a live examination. That afternoon when I walked into my apartment, the light on my answering machine was blinking. Pressing Play, I heard the voice of Roberta, our faculty adviser. Our model had just gotten her period, she began. It is possible to do a pelvic exam during someone's period, but it can be messy, and the model wanted to bow out. We needed to come up with someone to pinch-hit quickly or we'd have to cancel the training. Then she made her query: Would I consider stepping in as a replacement?

I weighed both sides. On one hand, I'd have strangers looking inside my vagina. I'd also have to see these strangers every week for the rest of the year. On the other hand, we'd have to deal with the pain of rescheduling and coordinating logistics for an alternate date.

It was not a hard decision. I called Roberta and said I'd do it. The next day I headed to the student health center and the awaiting stirrups of the exam table. There were several women whose faces I can't recall, but I remember the two men in the room—Niles, tall and lanky with light brown hair falling in his eyes, and Luis, compact with dark stubble and a dimple in his chin. Luis and I would go on a few dates later that year, but this exam was the closest he would ever get to my vagina.

You might think a pelvic exam in front of a live audience would make an impression, but I don't remember much—what the doctor looked like, what was said in the course of the instruction, or even how it felt. From prior experience, I know that the doctor pushed a metal speculum inside and opened it wide to reveal the vaginal walls and cervix. Here's where my memory finally kicks in. After everyone approached and poked their heads in close to see, the doctor sat me up, put a handheld mirror between my legs, and showed me my cervix for the first time.

Now I'm not bad looking, but my cervix is much cuter than the rest of me—a perfect fleshy doughnut with a small hole that could expand by tenfold to accommodate a baby's head someday. My eyes widened, and I shook my head and smiled. Miraculous.

For a few moments, I remained upright and chatted with the students and the doctor. Before I could ease myself down onto my back, Niles let out a sigh and crumpled to the floor. I had never seen anyone faint before, and it threw me off guard. Everyone else rushed to his side (I was still perched in a compromising position). Several sets of hands found his head and reached under his armpits to help him to his feet. He rose slowly, blushing and shaking his head.

Niles made it through the rest of that day and went on to be a kind and sensitive sex educator. Later, he and I would laugh together about how the power of my cervix once knocked him out cold. I lost touch with him, but I'd like to think he's a die-hard gynecologist out there somewhere.

These days, there are few opportunities for a twenty-year-old woman to show off her cervix to the world. At the time of my debut as a live pelvic model, I had already undergone three other Pap tests to screen for cervical cancer, at ages eighteen, nineteen, and twenty—I was quite a pro at them. Today, Pap tests aren't recommended until age twenty-one, then performed every three years until a person reaches age thirty.

After that, it's a bit of a choose-your-own-adventure; some clinicians choose to do a Pap test alone, an HPV test alone, or a combination of both. As opposed to testing every year, screening is now performed every three to five years depending on the method. Over the past two decades, we've realized that some abnormal Pap tests and even pre-cancers are transient. The immune system may resolve them without any intervention, but this process may take a few years. If you peek too often, you may overreact, treating an issue that might resolve on its own.

If a patient never develops pre-cancer or cancer, they can graduate from screening at the age of sixty-five. There is no award given at graduation, but if you've ever had a Pap, you'd agree that a certificate of achievement is in order.

When it comes to cancer prevention, you'd be hard pressed to find a greater success story than the Pap smear. At the time when George Papanicolaou developed the test in 1928, cervical cancer was the leading cause of cancer death among women. Because it was usually detected at advanced stages, the diagnosis was a death sentence. Radiation treatment was commonly used, but it caused pain, bleeding, infertility, skin burns; it bought someone more time but often didn't provide a cure.

Papanicolaou's test, a simple scraping of cells from the surface of the cervix smeared onto a glass slide, would eventually revolutionize the field of cervical cancer screening. It allowed for detection of pre-cancers, which could be treated early, reducing or eliminating the risk of cancer down the line. Papanicolaou's discovery would not catch on in earnest until the 1950s, when widespread roll out of the Pap smear caused cervical cancer incidence and mortality to plummet in the United States.[1] Today, cervical cancer has fallen from the first to the fourteenth most common cancer among women.[2]

But for all its successes, the Pap smear wasn't perfect. It was prone to giving false-negative results, misleading women with pre-cancer to think everything was fine. To compensate for the lack of sensitivity of the Pap test, clinicians performed them every year. Even if one test missed the diagnosis, hopefully the lesion would be caught during the next go-around, before it had a chance to develop into cancer.

Before the 1990s, anyone with a cervix had to resign themselves to a lifetime of annual Pap tests. These usually began in the teen years and continued every year ad infinitum, regardless of age. Once an eighty-year-old patient looked at me quizzically when I asked her to undress for her Pap. "There's been nothing going on down there for thirty years. Is this really necessary?" At the time, I insisted that it was. Most patients didn't question the annual routine. In everyone's minds, when it came to cancer screening, more must be better.

Meanwhile, as the medical community embraced the annual Pap smear, scientists debated over what was causing cervical cancer in the first place. Unlike other cancers, cervical cancer behaved very much like a sexually transmitted infection. It only occurred in women who were sexually experienced; the earlier someone started having sex and the more partners someone had, the higher the risk. Nuns and other celibates were almost never afflicted. By the 1970s, scientists were still stumbling around in the dark searching for the culprit. There were numerous suspects, but consensus at the time was that herpes simplex virus was most likely at fault.

German virologist Harald zur Hausen was skeptical about this theory. Zur Hausen certainly knew that viruses could cause cancer; he'd spent years studying the Epstein-Barr virus, the cause of mononucleosis (the "kissing disease"), and cancers of the nose and throat, as well as certain types of lymphoma. Zur Hausen believed that another family of viruses, the human papillomavirus (HPV), might be at the root of cervical cancer. At the time, HPV was known to cause warts on the feet and the genitals, but cervical cancer? This was a radical idea.

To build his case, zur Hausen would need to refute the current dogma, that herpes simplex virus was the cause of cervical cancer. In 1974, he headed to an international conference in Florida to present results of a study, where he'd found that herpes simplex virus was absent in cervical cancer tissue.

Shortly before zur Hausen's presentation, another researcher from Chicago took the wind out of his sails; he'd isolated 40 percent of the herpes simplex virus genome in a cervical cancer tissue specimen. In a profile of zur Hausen for *Cancer World*, Peter McIntyre describes the response to zur Hausen's presentation that followed: "The audience listened to zur Hausen in stony silence and dismissed his (now vindicated) results as lacking sensitivity. It was the low point of his professional life."[3]

Less than four years later, in 1977, zur Hausen's team would discover two new HPV types, numbered 16 and 18, determining that they caused nearly 70 percent of cervical cancers. HPV 16 was particularly powerful at causing cancer, turning up in a large proportion of cancers of the vulva, vagina, anus, and throat. Zur Hausen's discovery of cancer-causing HPV strains would earn him the Nobel Prize in Medicine in 2008. As for the herpes simplex virus, once falsely accused of causing cervical cancer? It was exonerated from that charge but still causes genital herpes, which is a hefty burden to bear in and of itself.

Not to discount the glory of the Nobel Prize, but this would not represent the pinnacle of fame for HPV. Little did zur Hausen know, but HPV would soon dethrone herpes as the most common STI, today infecting almost every sexually active person in the world. Then one day, HPV would infect just the right person, who would take zur Hausen's virus from a lowly laboratory bench straight to the stages of Hollywood.

All That Glitters

In today's vitriolic political climate, there is almost nothing we can all agree upon: gun control, reproductive rights, immigration policy, health care reform. But now and then, something comes along that unites us all, where we can all join hands and sing, "Kumbaya," regardless of race, gender, sexual orientation, or political affiliation. One of those great uniters is HPV.

In my field, we refer to HPV as the common cold of the genitals (thankfully, no sneezing is involved). According to the CDC, 85 percent of women and 91 percent of men will acquire HPV after having at least one sex partner, seventy-nine million Americans are currently infected, and another fourteen million will contract HPV in the next year.[4] No other STI has so many flavors: more than two hundred types of HPV have been classified, and about forty types can infect the anus and genitals.

HPV is an equal-opportunity STI, an unavoidable consequence of being sexually active. It infects everyone from baristas to high school teachers, professional athletes, even Hollywood stars. In her 2016 Netflix special, *Baby Cobra*, comedian Ali Wong boldly informed a live audience and millions of viewers that HPV had taken up residence inside her genital tract. And not to worry, she reassured the crowd, she wasn't the only one:

Everybody has HPV, okay? Everybody has it. It's okay. Come out already. Everybody has it. If you don't have it yet, you go and get it. You go and get it. It's coming . . .

A lot of men don't know that they have HPV, because it's undetectable in men. It's really fucked up. HPV is a ghost that lives inside men's bodies and says, "Boo!" in women's bodies.

My doctor told me I have one of two strains of HPV. Either I have the kind that's gonna turn into cervical cancer . . . or I have the kind where my body will heal itself. Very helpful, this doctor, right? So, basically, either I'm gonna die . . . or you're in the presence of Wolverine, bitches. We'll find out.

Most of the time, HPV doesn't cause a fuss. It just hangs out silently for years, not arousing attention while being passed from person to person. But as Wong alluded to, some HPV types are oncogenic: bad actors that can cause pre-cancer and cancer of the genital tract (cervix, vulva, vagina, penis), the anus, or the throat. Other low-risk types don't cause cancer, but they can cause warts in the anus and the genitals. And one or more HPV types can coexist in the same person at any given time.[5]

One thing Wong may not have been aware of: HPV is like glitter—it gets into everything. Once the cervix or the penis is infected, HPV can often be detected around the outside of the genitals and inside the anus, even without having anal sex. This glitter-like tendency to disperse is called a *field effect*. It may spread from the unavoidable rubbing of skin that occurs during sex, spread from infected fluids from one orifice to another, or spread from partial penetration (the "just the tip" phenomenon) that might occur if a penis or sex toy is involved.

For most women, HPV is an easy-on, easy-off infection. After having vaginal intercourse with a male partner, about half of teen girls will end up with HPV in the cervix within three years. There is a large range in time to the first infection, but Stuart Collins at the University of Manchester found that the average time was only three months.[6] If you flip this study around and take women with HPV in the cervix and test them repeatedly over time, about 90 percent of them will clear HPV within two years.[7] There's a lot of this on-again-off-again occurrence of HPV in the teens and twenties, when a quarter to one-half of women have HPV at any given time.[8] By the time women hit their thirties and beyond, their rate of HPV infection progressively declines.

With men, things are quite different. In the HPV Infection in Men (HIM) Study, Anna Giuliano at the Moffitt Cancer Center followed more than four thousand men aged eighteen to seventy years for four years, trying to see how HPV behaved in the male anogenital tract. To capture any traces of HPV "glitter," Giuliano was exhaustive in her sampling, covering the penis, scrotum, perineum, and around the anus. Not only was HPV infection more common among men than women, the proportion who were HPV positive didn't change much as men aged. Even men in their forties, fifties, and sixties had high rates of infection. Luckily, most men also cleared HPV and didn't develop genital warts or pre-cancers of the penis.[9]

If it's clear that HPV acquisition and clearance differs by gender, what does this mean for transgender people? Many transgender men and women take hormones of the opposite sex that they were assigned at birth, and it's unclear how this unique hormonal environment might play into HPV's comings and goings. To answer such a question would take hundreds of patients and several years. No one has done it yet, but I'm hopeful that someday soon someone will.

No matter how you look at it, HPV is incredibly common, so much so that it's unavoidable. Condoms protect somewhat but can't possibly cover all areas where HPV could be lurking, and a full-body condom seems a bit extreme. It's best to think of HPV as part and parcel of sex, something we all must deal with at some point. This is true no matter who we have sex with, even when it's just between ourselves and a battery-powered friend.

Give and Take

In the *Sex and the City* episode "The Turtle and the Hare," Miranda introduces Carrie and Charlotte to her new favorite sex toy, the Rabbit, a nine-inch-long pink vibrator composed of a bunny perched on a phallic shaft. Once inserted inside the vagina, the bunny's ears are perfectly positioned to provide clitoral stimulation while the shaft twirls and vibrates inside.

Despite some initial squeamishness, Charlotte soon develops a strong attachment to the Rabbit, becoming increasingly reclusive and forgoing plans with her friends to stay home with her new toy. An intervention is staged, and the vibrator is confiscated. Charlotte must return to navigating sex with other human beings.

After the airing of "The Turtle and the Hare," the Rabbit vibrator became an overnight sensation. In an interview for *Forbes*, Carol Queen of San Francisco's sex-toy store Good Vibrations recalled the day after the episode aired. She showed up for work and was greeted by "a line of women waiting to come into the store to check out the Rabbit. I'm pretty sure we sold out of them that day. We had people asking for the Rabbit by name." Vibratex, the maker of the Rabbit, saw its sales increase by 700 percent in the years following the airing of the episode.[10] More than two decades after its fifteen minutes of fame on *Sex and the City*, the Rabbit vibrator is still going strong.

The Rabbit also has a lesser-known claim to fame. It has served as an important tool of scientific discovery in the field of HPV transmission. It all started when Teresa (Anderson) Batteiger at Indiana University became curious about whether sex toys were potentially a source of HPV transmission between sex partners, particularly women who have sex with women. It was a reasonable question. In the 1990s, scientists had detected HPV on people's underwear and other inanimate objects. Others had found that HPV could survive and be infectious after drying in plastic tubes for up to a week.[11] Why wouldn't the same be true for sex toys?

Batteiger recruited twenty women to test out her theory. She supplied them with two vibrators: vibrator 1 was a "Rabbit-styled" vibrator made of a jelly-based thermoplastic elastomer; vibrator 2 was a smooth, soft silicone. After making a pilgrimage to Good Vibrations myself, I can report that both are smooth, but the Rabbit is a little softer and more pliable to the touch. The thermoplastic jelly of the Rabbit is also more porous than silicone, with more nooks and crannies where bacteria and viruses might hide.

In addition to the vibrators, Batteiger supplied participants with a commercial sex-toy cleaning product, and long Q-tip swabs. After swabbing the inside of their vaginas to test for HPV, the participants were asked to use each vibrator inside the vagina (without a partner), separated by at least twenty-four hours. After using the vibrators, Batteiger asked the women to swab the surface of the vibrator on three separate occasions: 1) immediately after use, 2) immediately after cleaning, and 3) twenty-four hours after cleaning.

Out of the twenty women that initially agreed to participate, twelve returned samples that could be tested for HPV. The other eight women took the vibrators and test kits but never returned any samples. In clinical research, we call these participants *lost to follow-up*, but it's easy to figure out where they might have gotten waylaid. We can only hope that someone eventually found them and pried the vibrators out of their hands.

Batteiger discovered that HPV was stubborn; it could be

detected immediately after the vibrators had been used *and* immediately after cleaning. Both types of vibrators held on to the HPV, but the plastic jelly Rabbit held on to it longer than the silicone model. Even twenty-four hours after cleaning, two out of five Rabbit vibrators were still positive for HPV.[12] The bottom line? Better not to borrow someone else's bunny; it might be riddled with HPV.

Even without penetration from sex toys or penises, HPV can easily be shared. External genital-to-genital contact can create plenty of friction for successful transmission of HPV. Then there's the controversy of the hands. If you have HPV in your genitals and you touch yourself, then it is certainly possible to detect HPV on your hands and under your fingernails. But can HPV on your hands successfully infect the genitals of your partner?

Over the course of twenty years, five different studies explored this, and the answers were as clear as mud: some concluded that it was possible, others concluded that it was unlikely, and some concluded that they couldn't conclude either way. In 2019, Talía Malagón from McGill University may have finally put the issue to rest. She tested hand and genital samples from hundreds of women and their male partners over time to see whether new HPV infections that appeared in the genitals could originate independently from the hand after accounting for genital contact. Her conclusion? Genital HPV infections are far more likely after genital penetration, rather than hand-to-genital contact.[13]

One disturbing aspect of Malagón's study was the abundance of HPV on people's hands: a third of her participants had HPV on their hands, even *after* handwashing. A patient with obsessive compulsive disorder (OCD) reached out to me when he heard this because he was seriously distressed—he imagined that everyone was walking around with HPV on their hands, touching surfaces in public restrooms that he'd have to touch like faucets and doorknobs. Luckily Malagón's study can reassure even the OCD among us. Yes, HPV is likely out there on surfaces, but as long as you don't stick that doorknob where it doesn't belong, sex is still the most likely way to catch HPV.

The science of HPV transmission between couples is much like sex itself: messy and often complicated. Researchers Linda Widdice and Anna-Barbara Moscicki at the University of California–San Francisco made an admirable attempt to sort out the mess in a couples' transmission study, which they refer to as *Sex on Sunday*. In the study they recruited twenty-five monogamous heterosexual couples, asked them to have sex on Sunday night, and then come to the clinic on Monday to be tested for HPV.[14] Widdice and Moscicki then looked everywhere they thought that HPV might be transmitted during sex: in or around the anus, the vagina, penis, scrotum, hands, mouth, and tongue.

Widdice and Moscicki found that 68 percent of couples shared genital HPV types with each other.[15] Shared HPV was less common if men washed their genitals after sex and if intercourse was spaced further apart. Notably, unlike every other STI and HIV, females were twice as likely to transmit HPV to males than the other way around.

When they tested the anus in both members of the couple, things were murkier. There was less HPV all around, and less agreement in types. Particularly when a man has sex with a woman, the pathway to an HPV anal infection is not direct: perhaps from her vagina, to her fingers, to his anus. To complicate matters, Widdice and Moscicki only tested around, not inside the anus of the men; they were concerned that men would refuse if they asked to stick a swab inside. To truly understand HPV in the anus would require an anally focused study.

And what kind of person would be interested in conducting

such a study? It turns out Moscicki had a colleague down the street who was already delving into the anus with gusto, but who was unsure whether his pursuits would lead to a dead end.

Rear Window

Since 2003, *Popular Science* magazine has published an annual list of the Worst Jobs in Science, which writer Jason Daley referred to as "our annual bottom-10 list, in which we salute the men and women who do what no salary can adequately reward."[16] I think the point of this list is to make its readership feel better about their own career choices. "Well, my job is bad, but at least I'm not doing *that*."

In 2004, at the top of the list was Joel Palefsky and his colleague Naomi Jay at UCSF. Their job, anal wart researcher, beat out worm parasitologist for the number-one Worst Job in Science. To my surprise and slight resentment, in the number-four spot that year was my job, STI researcher, also known as *tampon squeezer* (apparently, tampons make great tools for collecting vaginal discharge for research).[17] Frankly, spending time in people's rear ends or studying STIs is much more appealing to me than some winners from past years, including barnyard masturbator, flatus odor judge, and dysentery stool sample analyzer. Clearly, it takes all kinds of people to run the world.

As his job title would imply, Palefsky is the scientific pioneer who first detailed the comings and goings of HPV infection in the anus, describing its relationship to both anal warts and anal cancer. I often describe him to other colleagues as the Elvis of anal HPV. This association with the King is due to the magnitude of his academic celebrity, not the tight white jumpsuit or the addiction issues.

He chuckled as he reflected on his unusual career path. "Nobody grows up aspiring to be an anal HPV person. Pardon the expression, but I really came at this through the back door." Palefsky was trained as both an infectious disease physician and virologist. Like many in the field of HPV, his research interests began in the cervix. But then his training brought him to San Francisco in 1989 at the height of the AIDS epidemic. Everywhere he turned, he saw anal HPV, showing up as severe cases of warts among gay men dying of AIDS.

Unlike the cervix, which was crowded with other researchers, no one was clamoring to get into the anus with Palefsky; it was uncharted territory. He knew that strains of HPV that caused cervical cancer could also cause anal cancer. In the general population, anal cancers were more common among women than men, but gay and bisexual men were at much higher risk than the general population. He wondered how HIV could play an additional role. Similar to cervical HPV, perhaps HIV's suppression of the immune system would prevent patients from clearing an anal HPV infection, fueling high rates of anal cancer among those who were HIV positive.

To find out, he launched a series of studies among HIV-positive and HIV-negative patients, which he refers to as the *Tushie Studies* (One through Five). In Tushie One, Two, and Four, Palefsky tested men who had sex with men for anal HPV and signs of anal pre-cancer by performing an anal Pap test.[18,19] To perform the anal Pap, he and his team took long Q-tips and inserted them into the anus, twirling them around to sample the cells at the junction of the anus and the rectum. Then following the Pap, they examined the inside of the anus with a technique called *high-resolution anoscopy*.

During the high-resolution anoscopy, a clear plastic tube was inserted in the anus to hold it open, and then the tissues were examined closely with a binocular microscope. Vinegar was liberally and repeatedly applied inside the anus to turn abnormal tissues white, which lent the exam room a faint odor of salad dressing. Any suspicious-looking tissue was biopsied to see whether it was pre-cancerous, and pre-cancers were treated and closely monitored.

I've only performed anoscopies and not had one myself, but from my vantage point, patients seem to tolerate the procedure well. When I asked a patient about discomfort, he replied that it was "not exactly fun, but better than the dentist." My mentor Michael Berry taught me to chat during the procedure to pass the time; this worked wonders at putting patients at ease. I had only one patient where chatting didn't sufficiently settle her nerves: she took too much Xanax, called her boyfriend, and tried to have phone sex with him as her distraction. I had to cut that off. While it was working for her, it was much too distracting for me.

During the Tushie Two study, Palefsky and his team performed almost three thousand of these anoscopies and anal Pap examinations on more than six hundred HIV-positive and HIV-negative men who had sex with men. What he found was astounding. Anal HPV was almost universal: nine out of ten HIV-positive men and six out of ten HIV-negative men had least one type of HPV in their anus. At least one-third of the HIV-positive men had HPV 16, the type responsible for the majority of anal cancers. Among the HIV-positive men, more than half of them had an abnormal anal Pap test that could indicate presence of an anal pre-cancer.[19,20]

When he turned his attention to HIV-positive women (of all sexual orientations) in Tushie Three, he found more of the same. In fact, when Palefsky first saw the women's results, he called the laboratory to see whether they had switched the labels on his samples. He found that HIV-positive women were *twice* as likely to have HPV in the anus as in the cervix.[21] He repeated this study later and added in HIV-negative women (Tushie Five). His results were similar. Even women who had never had anal sex still had anal HPV, and lots of it.

Though Palefsky was uncovering all this anal HPV, anal cancer was still rare, even among patients with AIDS. Before 1996, AIDS patients were dying of opportunistic infections such as *Pneumocystis* pneumonia—they weren't living long enough to develop anal cancer. When antiretroviral therapy for HIV came along in 1996, everyone hoped that restoring the immune system would also allow patients to clear anal HPV, bringing their anal cancer risk back to baseline. Palefsky worried the opposite, that a longer life span would allow HIV-positive patients to live long enough to develop anal cancer in droves.

Unfortunately, Palefsky was right. After 1996, people lived longer with HIV, but they developed diseases of aging, such as heart disease and cancer, often at higher rates than their HIV-negative counterparts. Palefsky gradually watched rates of anal cancer tick upward for HIV-positive women and men. By 2012, researcher Michael Silverberg found that anal cancer was twenty times more common for HIV-positive women and heterosexual men, and up to eighty times higher for gay and bisexual men with HIV.[22]

If a cancer were running rampant in certain segments of the population and a screening test existed that might prevent it, it seems logical that people would start swabbing away at the anus to try to detect cancer. That's not what happened. Unlike the cervical Pap test, Palefsky didn't have definitive evidence that performing an anal Pap and treating pre-cancer would prevent anal cancer. It seemed like common sense, but common sense alone was not enough to sway national organizations who created clinical guidelines. Palefsky sat in national meetings and found himself agreeing that there wasn't enough evidence to recommend the anal Pap to high-risk patients, even though he felt like it was the right thing to do.

To make matters worse, after ten years of funding, the NIH pulled the plug on his Tushie studies in men. In his understated way, he expressed his disappointment—"Well, I certainly wasn't pleased"—but he wasn't sore at the NIH. "I always take responsibility myself," he said. "I didn't make it clear enough, or I simply wasn't able to get them sufficiently excited about that area of study. So, I started thinking, what is the next most important thing to address?"

He set his sights on conducting a large clinical trial, one that would answer the question about whether screening with an anal Pap and treating pre-cancer would lower the risk of anal cancer. He toyed around with potential names. First, he thought of the Anal Screening Study (ASS), but that was a little too vulgar. He settled on ANCHOR, the Anal Cancer HSIL (high-grade pre-cancer) Outcomes Research Study.

He called up the National Cancer Institute (NCI) at the NIH to discuss his idea—a multicenter, multimillion-dollar anal cancer screening study, requiring at least five years of follow-up for each patient. The NCI was interested, but funding such a study would require a special financial agreement, and Palefsky would need to jump through many bureaucratic hoops before they could proceed.

It only took ten years of negotiating. Then in the spring of 2015, Palefsky was back at his family home in Montreal when his cell phone rang. On the other end was his colleague at the NCI. His proposal had passed their last hurdle, and the NCI was going to enter into a multiyear agreement with him, to the tune of $89 million.

After hanging up with his colleague in Bethesda, Palefsky

came back to the table and announced the news to his family. There was no shouting or popping of champagne corks (these are Canadians, after all). Someone patted him on the back, said, "Good job," and then everyone resumed eating their dinner.

An ad for Palefsky's ANCHOR Study, which will help determine whether anal Pap testing and treatment of pre-cancer will prevent anal cancer. Study supported by the National Cancer Institute of the NIH under U01CA121947. Content is solely the responsibility of the author and does not necessarily represent the official views of the NIH.

Unless any safety concerns arise, ANCHOR will be going on until 2026. In the meantime, Australian investigators in SPANC (Study for the Prevention of Anal Cancer) decided to watch almost two hundred men with anal pre-cancer without treatment for three years to see what happened. Only one new cancer developed; most of the pre-cancers just sat there, and one in five of them regressed on their own.[23] Their findings beg the question: Should we treat anal pre-cancer if it's going to get better on its own? Probably not—but which pre-cancers are transient and which ones will progress to cancer? Hopefully, the ANCHOR study will clear up some of these unanswered questions.

Until ANCHOR is complete, it is unlikely any national organizations will recommend anal Pap testing. For now, to detect anal cancer, an old-fashioned finger in the bottom is the best thing we have. Palefsky touts this exam as he travels the world proselytizing about anal cancer awareness. He's sometimes frustrated that "doctors and other clinicians are not putting their fingers in to examine the anus or looking at it, in part because of the stigma around this body part." The unfortunate truth is that most doctors don't even feel comfortable talking about anal sex, much less putting their finger in someone's bottom.

Stigma around anal cancer might be one of the last remaining cancer taboos. Many others have been shattered, due in part to celebrities coming out and sharing their personal battles. Even cancer of intimate organs is fair game: Lance Armstrong's testicles, Angelina Jolie's breasts, Ben Stiller's prostate. All these stars have come out publicly about their illnesses, creating a community of public support rather than stigma or shame. Palefsky laments that "it's hard to get well-known people to support the cause. We don't have big-name star-studded galas for anal cancer."

In 2006, Farrah Fawcett became the first celebrity to go public with her anal cancer diagnosis, filming a video diary of each grueling step of her battle, which she lost in 2009. Two years after Fawcett's death, Palefsky was aghast as Ryan O'Neal spoke about her cancer on *Piers Morgan Tonight,* telling Morgan that toxic stress from O'Neal and his children had caused her cancer. "If she had never met us, would she still be alive today? Because nobody knows what causes cancer, do they, really?" (Perhaps O'Neal was unaware or chose not to mention that 90 percent of anal cancer is caused by HPV.) Morgan queried, "Do you really believe that?" to which O'Neal replied, "I think it's highly possible."[24]

Palefsky bristled at these remarks. "All we needed to know was that it was a sexually transmissible virus that caused her cancer. And that it was preventable with vaccination. A lot of lives could have been saved if there had been a clearer connection made to HPV. We don't have to talk about anal sex—you don't need to have anal sex to get anal HPV. Whether it gets directly to the anus or is spread there from the cervix, it doesn't matter."

In September 2018, *Desperate Housewives* star Marcia Cross posted about her cancer-related hair loss on Instagram, letting everyone know that she was fighting anal cancer. In a March 2019 interview for *People,* she announced she was in remission. Her intentions with the announcement were clear. "I want to help put a dent in the stigma around anal cancer. I've read a lot of a cancer-survivor stories, and many people, women especially, were too embarrassed to say what kind of cancer they had. There is a lot of shame about it. I want that to stop."[25]

Cross is now an enthusiastic anal cheerleader. "Having woken up to its importance, I am now a big fan of the anus!" Unfortunately, *People* didn't mention HPV as the cause of anal cancer, but dozens of other news outlets picked up her story and made the obvious connection. And the good news is that Cross's and Farrah Fawcett's anal cancer may one day be a thing of the past. Just as Fawcett was receiving her diagnosis, scientists in Australia were celebrating their role in the development of an effective vaccine against HPV, one that could help prevent cancers of the throat, genitals, and anus. In under a decade, this vaccine would have an incredible impact on HPV-related disease on the Australian continent, its success due in part to a quite unexpected supporter.

The Perfect Storm

On October 16, 2006, at the Pink Ribbon Breakfast for the Australian National Breast Cancer Centre, Janette Howard, the prime minister's wife, was discussing the good news about advances in breast cancer treatment and survival. It was common knowledge among Australians that Howard was a breast cancer survivor, and her cancer had been discussed by all the national news outlets at the time of her diagnosis.

Once in recovery, she became a patron of the National Breast Cancer Centre and an advocate for screening and early detection programs. But in ten years, she had never discussed her personal experience with the disease. She was intensely private and routinely sidestepped media interviews and public appearances.

As Howard stood at the podium discussing the strides made in breast cancer survival, she announced to the crowd, "I should perhaps take this opportunity to say [that] I didn't have breast cancer—my cancer was cervical cancer." Then she added, "And there are, you know, media stories that I perhaps at this point ought to sort of clear up."[26]

In today's age of oversharing, where celebrities like Bella Thorne live tweet their colonoscopies, such a statement would barely cause a ripple. But from a politician's spouse who never spoke of her personal life, this was a newsworthy revelation. Multiple outlets all over Australia publicized the news, placing the Australian public's attention on cervical cancer in a way that it never had before.

As luck would have it, just four months prior to Howard's

speech, the pharmaceutical giant Merck had announced the release of a highly effective quadrivalent vaccine that would protect against four types of HPV: types 16 and 18, which caused about 70 percent of cervical cancers, and types 6 and 11, which caused about 90 percent of genital warts. The technology behind the vaccine had been developed in Australia by immunologist Ian Frazer and a Chinese molecular virologist named Jian Zhou.

Unlike vaccines for diseases such as measles or polio, the HPV vaccine did not contain any live or killed parts of the virus. It was composed of a protein called *L1* from the HPV viral capsid, or the virus's outer protein shell. Frazer and Zhou had used insect cells to produce the L1 protein, and Zhou had figured out how to make the protein self-assemble into a viruslike particle that looked identical to the actual virus. The viruslike particle was empty on the inside, but successfully fooled the immune system into thinking that it was seeing HPV, and in response, it would produce high levels of protective antibodies, much higher than those obtained from a natural infection.

In 1991, Frazer and Zhou first published their discoveries, which they shared with pharmaceutical companies CSL, Merck, and GlaxoSmithKline. Frazer advised them, "If you're going to have a vaccine (against HPV), this will be where it comes from."[27] Then in 1999, Zhou passed away of a mysterious illness, leaving Frazer to continue working solo. Frazer felt confident that they were onto something with the viruslike particles, but he couldn't be sure that it would see the light of day. The process of vaccine development often takes many years from the basic science work to actual product release. Most vaccine candidates fail, and vaccinologists don't often see the fruits of their labor.

In June 2006, the release of a cervical cancer vaccine by Merck based on Frazer and Zhou's viruslike particles set off a firestorm of excitement. A second HPV vaccine based on the viruslike particle was released by GlaxoSmithKline eleven months later. Frazer was reluctantly catapulted into stardom, winning numerous awards, including Australian of the Year, which came with a similar amount of cachet as *People* magazine's Sexiest Man Alive award might afford in the United States. (Frazer was not a candidate for *People*'s award, but may I say, he is a bit of a dish.) Along with Frazer's newfound fame and Janette Howard public status as a cervical cancer survivor, HPV was thrust into the Australian public's consciousness, and the time was ripe for the country to act.

In 2007, Australia became one of the first countries in the world to launch a nationally funded HPV vaccination program, providing free school-based vaccination to all twelve- and thirteen-year-old girls. The government also provided free HPV vaccines to physician offices all over the country so that teens and young women up to the age of twenty-six could also be vaccinated. Although cervical cancer prevention in women was the initial goal, in 2013, they added boys to the national vaccination program. Today, Australia's HPV vaccination program is admired the world over; 76–80 percent of fifteen-year-olds have received three doses of the HPV vaccine.[28]

The result is that HPV and associated diseases are beginning to disappear in Australia. There has been a 77 percent reduction in cancer-causing HPV 16 and 18, and pre-cancers in girls have also decreased by 50 percent. If current trends continue, by the year 2030, Australia will have fewer than four cases of cervical cancer per one hundred thousand women per year, which is below the threshold for cervical cancer elimination set by the World Health Organization.[29]

The HPV vaccination program has also led to great success in Australia in the takedown of the genital wart. At the International Papillomavirus Conference in 2018, Eric Chow of Monash University announced that genital warts were now a rare sighting in the country: there had been a 90 percent reduction in cases of genital warts among fifteen- to twenty-year-old Australians.[30] Warts were only flourishing in the genitals of the over-thirty-six crowd, who were not age-eligible for vaccination under the national program.

As Australian boys and girls age into eligibility for the national vaccination program, the proportion of the population vaccinated against HPV-related genital warts (and cancers) will continue to grow. Over time, when a critical mass of the sexually active population is vaccinated, warts will likely decline across the population, even among people who aren't vaccinated, a concept known as *herd immunity*.

My favorite example of the power of herd immunity comes from Christopher "Kit" Fairley from Monash University in Melbourne. Say a 747 aircraft full of lustful American teenagers with genital warts lands in Australia. An average 747 plane with a typical three-class layout (coach, business, first) can hold 416 people. Everyone deplanes and starts having sex with an Australian teenager during the next six months. Would this cause enough new cases of genital warts to spread throughout the population once again?

According to Fairley, no. Genital warts can only survive in Australia if each person with HPV-related warts can transmit the infection to at least one other individual. Today, the likelihood of someone successfully transmitting HPV to an Australian teenager is low: 80 percent of them are vaccinated against HPV, and the likelihood of transmitting HPV to people who are vaccinated is nearly zero.

What if a passenger has sex with an Australian who isn't vaccinated? According to a study by Ann Burchell from the University of Toronto, about 20 percent of uninfected people who are exposed to HPV will be infected within six months.[31] Not everyone who contracts HPV will develop a genital wart, but for this example, let's assume that they do. So only 20 percent of the adolescent population is unvaccinated and vulnerable to infection, and only 20 percent of those will actually catch an infection if exposed. After the 416 Americans with warts deplane and have sex with one partner, they will successfully spread HPV to about 16 other people, far less than the 1:1 ratio needed to maintain the spread of warts in the population.

Based on Fairley's example, the genital wart does not stand a chance against the strength of herd immunity in the Australian population. If current vaccination rates hold steady, we are unlikely to hear much about genital warts in Australia in the future. I don't think anyone will miss them much.

Despite lagging behind Australia in adolescent HPV vaccination (half of U.S. adolescents are vaccinated compared to about 80 percent in Australia),[32] there are also positive signs of vaccine impact here at home. In 2017, the CDC reported that infections with HPV vaccine types had decreased by 71 percent among fourteen- to nineteen-year-old females.[33] In 2019, Nancy McClung at the CDC estimated that cervical pre-cancers in the United States had declined among eighteen- to twenty-four-year-old women from 2008 to 2016 (216,000 cases to 196,000 cases), the first glimpses that prevention of cervical cancer may be a reality.[34]

Delays in scaling up HPV vaccination in the United States are rooted in multiple controversies, which have surrounded the vaccine from day one. Shortly after the vaccine's release in 2006, there was aggressive marketing and lobbying of politicians by Merck to promote it as a mandatory vaccine, causing a backlash and public mistrust. As a new vaccine, there were concerns about whether it would prove safe when administered widely outside of a clinical trial. Fourteen years and more than one hundred million doses later, the evidence is overwhelming that the HPV vaccine is safe and effective. So why are there some who are reluctant to use it?

Imposter Syndrome

On April 30, 2018, Lars Andersson from the Department of Physiology and Pharmacology at the Karolinska Institutet, Sweden's largest academic medical center, published an alarming editorial commentary in the *Indian Journal of Medical Ethics* on the impacts of HPV vaccination in Sweden.

After analyzing data from the nation's Centre for Cervical Cancer Prevention, Andersson noted that there had been increases in cervical cancer rates among women twenty to forty-nine years of age in Sweden. He argued that the HPV vaccine might be *causing* cervical cancer if given to women who'd already been exposed to HPV, instead of preventing it.[35]

It was a disturbing conclusion, and a curious one, given that most of the women under study were not age-eligible to be vaccinated. Even among women twenty to twenty-three years of age (the only age group who might have been vaccinated during the study period), it was unknown whether the women with cancer in that age group had actually received the vaccine.

This was not Andersson's first editorial about a vaccine's potential pitfalls. Between 2016 and 2018, he had published other editorials highlighting his theories about the links between vaccines and the onset of autoimmune diseases, focusing on the HPV vaccine and chronic fatigue syndrome, and the swine flu vaccine and type 1 diabetes. Upon reading Andersson's opinion pieces, some scientists criticized his analysis methods as incorrect or overly simplistic. Still, some reputable journals, including

Vaccine and the *Journal of Internal Medicine,* felt his opinions were worth publishing in the interest of scientific discourse.

There was, however, one small problem. Lars Andersson was a fake. There was no one by that name in the Department of Physiology and Pharmacology at the Karolinska Institutet.

After the *Indian Journal of Medical Ethics* learned of the deception, Andersson's affiliation was removed from the commentary, but the journal did not initially retract the article. Instead, they reached out to the author who "informed us that he had used a pseudonym besides a false affiliation." He revealed his identity to the journal's editor, who determined that he did "face a credible threat of harm" because of his unpopular opinions and thus felt the journal ought to keep his identity private. It was only after watchdog blogs and critics barraged the journal with complaints that the journal retracted the paper.

But the damage from "Lars Andersson" had already been done. While Andersson's findings were not widely cited by other scientists due to lack of merit, they were cited by bloggers and anti-vaccine organizations who used his publications as "evidence" for claims of harm. That fake news could be published as fact revealed fatal flaws in the practice of medical publishing— and anti-vaxxers could then use these weaknesses to their advantage.

When searching for facts online, distinguishing anti-vaccine pseudoscience from real science can be nearly impossible. If a parent searches online for any vaccine, anti-vaccine blogs and websites with official sounding names (e.g., National Vaccine Information Center) will appear, seeming just as reputable as mainstream medical or public health sites. Some are written by people trying to peddle their own "natural" remedies, others by highly credentialed experts rallying against the establishment. Common themes include conspiracy between Big Pharma and the medical establishment, and skepticism for science—except when there is any evidence demonstrating harm from a vaccine. (If this "evidence" has been discredited or retracted, that fact may be conveniently absent.)

Not everyone is inclined to believe anti-vaccine propaganda, but there are certainly people who do. In a study of over one thousand U.S. adults, Matt Motta from the University of Pennsylvania found that more than a third of participants thought they knew as much or more than scientists about autism, including beliefs that autism is caused by vaccines. This overconfidence was associated with low knowledge scores on a ten-item questionnaire and support for nonexperts (e.g., celebrities) playing a role in policy making.[36] Overconfidence in one's knowledge is known as the Dunning-Kruger effect, or "Ignorance of my own ignorance," and may be one driver of public anti-vaccine sentiment. (This is the opposite of the imposter syndrome: "I know what I'm talking about, but I'm afraid people think that I don't.")

Other sites use the power of narrative to convince people of vaccines' harms: tales of children or teens who became sick at some point after vaccination, developing autism, chronic pain, lupus, rare cancers, infertility. Every one of these stories is devastating. I ask myself what it would be like if my child was severely ill. Wouldn't I want to determine the cause of the suffering?

The fact that a parent *thinks* their child's illness is caused by a vaccine is enough to plant fear in another parent's mind. Cornelia Betsch from the University of Erfurt studied parents' attitudes toward vaccination after browsing vaccine-critical websites versus neutral government education websites. A mere five to ten minutes of browsing was enough to change parents' perceptions: that vaccination was more dangerous to them than before, and the risks of not vaccinating were decreased in their minds. Five months later, parents who perceived higher risks after browsing were more likely to search for other vaccine-critical information and were less likely to have vaccinated their kids with recommended vaccines.[37]

What was the main difference between websites? The antivaccine websites told more personal stories than the government website—changing hearts and minds, sometimes irreversibly.

Here's the issue, which any doctor or clinician can tell you firsthand: terrible stories happen every day; all forms of medical catastrophe can and do occur. Children and young people develop autoimmune diseases and other debilitating illnesses out of the blue, and there are even instances of unexplained sudden death. Some of these issues will develop at some point (days, weeks, months) after receiving a vaccine. If these outcomes don't happen more often to vaccinated people than unvaccinated people, then it is incorrect to blame the vaccine for the bad outcome; it might have happened anyway.

Clinicians have stories of our own, and we need to spend more time telling them. I've had patients in their teens who were raped and then developed severe pre-cancer, women in their twenties who've undergone multiple surgeries of the cervix, women in their thirties who lost healthy pregnancies after these surgeries. These outcomes were certainly the result of HPV and could likely have been prevented with vaccination.

The battle between antivaxxers and scientists around the HPV vaccine has always been fierce. First, there were claims that vaccinated teens would be more prone to risky sexual behavior (not true).[38] There were Republican presidential candidates claiming that the HPV vaccine caused mental retardation (also not true). Then there were legitimate questions about links between vaccination and rare but terrible outcomes. Scientists and the CDC have actively searched for links between HPV vaccine and premature ovarian failure, chronic regional pain syndrome, postural orthostatic hypotension (sudden drop in blood pressure when standing), birth defects, pregnancy loss, the list goes on.[39,40] After the 100 million doses distributed in the United States, and 270 million doses worldwide, there have still not been any associations found. Public health practitioners are certainly willing to acknowledge bad outcomes if they exist, but we're still waiting to see some proof.

As I prepare to vaccinate my own son against HPV, I find there is little explaining I have to do. He has grown up looking over my shoulder as I prepared my lectures and leafed through my textbooks. "Everybody gets HPV," I told him, "and someday you will too." And before I knew it, he had. At age nine he developed a common wart on his knee. As his doctor was freezing it off, she explained that it was caused by HPV. He shrieked with delight, walked into the hallway, and announced, "I have HPV!" setting off a twitter of giggles from passersby. While I'm glad he's okay with it now, I don't think he'd be as happy if HPV were to wreak havoc in his genitals. And once he's had the HPV vaccine, it's not likely that it will.

5

Affectionate and Popular

Oh, the Complex Sexual Webs We Weave

The Tale of William and Xavier

William had driven the entire eight hours from Las Vegas to San Francisco's City Clinic with his "penis on fire." He announced this as I ushered him back to the examination room, closing the door behind him. He sat and shifted uncomfortably while removing his hat, revealing dozens of fresh cornrows in an intricate geometric pattern. I complimented his braids, which elicited a smile, his flaming genitals momentarily forgotten.

William was in his midtwenties and had been living between San Francisco and Las Vegas, trying to make it as a choreographer. Life in Las Vegas was great, except he was pretty sure he had contracted gonorrhea from his new girlfriend. If he was correct, it would make this his fifth episode in just two years. He furrowed his brow and gestured with both hands as if to say, "What gives?" In his estimation, he was not a high-risk guy. In fact, he considered himself monogamous, at least in a geographic sense. When he was in San Francisco, he had sex with one woman, and in Las Vegas, he had sex with another one. He had been back and forth every few weeks but had only slept with these two women in the past three months.

He had started off using condoms with his partner in Las Vegas, but she was taking birth control pills, so he had stopped using them in recent weeks. A week after they had slept together the most recent time, he started to feel aches and pains, what he referred to as "jolts of lightning" down in his pelvis. After four prior episodes of gonorrhea, William was so attuned to his body that he could tell immediately something was wrong. He sought help at a local clinic, but they wanted eighty dollars from him just to walk in the door, plus more if he needed treatment.

Being cash-strapped at that moment, William went home without being seen. After sleeping on it for a night, the discharge from his penis began in earnest. He remembered the free STI clinic in San Francisco and suddenly realized what he needed to do. He put on clean underwear, hopped in his car, and headed west.

He arrived in San Francisco early on Good Friday and headed straight for City Clinic in the rapidly gentrifying neighborhood south of Market Street. Inside, the crowd was light, and shortly after arriving, he found himself in a room, lowering his pants for me, a complete stranger.

A yellowish fluid oozed from the tip of his penis. Whether referred to by its common name of *pus,* or the more official *purulent discharge,* neither one had any business coming out of his penis. I gently wiped off the fluid with a long swab and spread it onto a glass slide. Then I inspected the rest of him quickly. Finding no other bumps, ulcers, or rashes, I took William's slide over to the laboratory while he waited in the room.

After fixing the specimen to the slide with a quick touch of heat, I applied dyes to prepare the slide for viewing under the microscope, a more than one-hundred-year-old process known as *Gram staining*.

Placing the slide on the stage of the microscope, I turned the light up to its brightest setting and focused my image. Even with low magnification, it was evident that William's immune system was fighting something. The slide was covered with hundreds of white blood cells, a marker of inflammation, the body's defense against foreign invaders like bacteria and viruses. Due to the Gram staining process, the white blood cells had stained pale pink, and the key to the diagnosis could be found within those cells. Inside at least a dozen of them were multiple pairs of hot-pink spherical bacteria: *Neisseria gonorrhoeae*.

An ancient disease, the moniker *gonorrhea* has been credited to the Greek physician Galen (A.D. 130–200) after *gonos* (semen) and *rhoia* (to flow). It became widely known as *the clap* before its true bacterial origin was identified, possibly from the French *les clapiers*, referring to Parisian huts where prostitutes serviced their clients. The nickname may also have arisen in reference to primitive treatments that involved clapping a hapless penis between two wooden paddles, pair of hands, or books in an attempt to expel unwanted discharge.[1]

Even in 1879 when Albert Neisser discovered the gonococcus bacterium, treatment options were similarly Draconian: heating metal probes to temperatures of over 110°F and inserting them into the vagina or rectum for hours or instilling the urethra with metallic compounds of silver and mercury.[2] So despite the unfortunate diagnosis, William was lucky to be getting the clap today, when antibiotics would cure him within days rather than having to endure weeks of suffering.

Walking back into the exam room, I sat down and faced him, stating simply, "You have gonorrhea again." I often apologize before I give someone an STI diagnosis, as if to say, "Sorry to be the bearer of bad news." Since I was only confirming what William had already suspected, an apology didn't seem appropriate. If anything, he looked satisfied. "I knew it. I told you, I know my body."

His theory about how it happened this time? "I'm having bad luck with sex, especially oral sex." That was definitely possible. Gonorrhea can certainly be transmitted to men via oral sex, although I couldn't prove how he had been infected or which of his partners was ultimately responsible.

Now that he knew his diagnosis, he seemed more relaxed. I offered an important reminder: "You know you'll feel better soon, but you and your penis are off limits for the next seven days." The week off from sexual activity would allow the antibiotics time to kill off the gonorrhea completely and prevent him from getting reinfected from one of his partners.

He nodded. "I swear I won't be doing anything with these females for a while."

William planned to tell each of his partners about having gonorrhea. It was tricky because he hadn't disclosed to either of them that he had been sleeping with the other one. He wasn't lying to them, but they never asked, and he didn't offer. In fact, he thought it was possible that *they* were also having sex with other guys when he wasn't around.

I passed him off to our nurse, who administered two antibiotics, an intramuscular shot in the arm plus four pills, which he took immediately. William had doubts that his partners would be able to access medical care quickly, so the nurse handed William more doses of antibiotics for them in two separate paper bags, gonorrhea treatment to go. We knew that expediting his partners' treatments this way was one of the most effective interventions for preventing repeated infections.[3]

If each of William's partners took their antibiotics, and if they all took a week off from having sex with each other or anyone else, the chain of infection would be broken, resulting in a happy end-ing for all. It's the frequent failure of those two ifs, though, that has fueled the need for our clinic's continued existence for more than seventy years.

William's tale seems straightforward at first glance. He had unprotected sex with two partners, and he contracted gonorrhea. What about the fact that he had suffered so many bouts of gonor-rhea over the past two years? William attributed it to bad luck, and a decade ago, I might have believed him. But the picture became a little more complicated when Xavier checked in later that day.

Like William, Xavier was also in his midtwenties, but he was Latino, gay, and distinctly *not* monogamous, geographically or otherwise. His dance card had been quite full recently. When I asked him how many sex partners he'd been with in the past three months, he looked up at the ceiling to help him concen-trate while he counted on his fingers. He went through both hands twice before he felt confident to answer: somewhere be-tween twenty and twenty-five. He didn't have any symptoms and was just coming in for an STI checkup, which was something he did religiously every three months.

He reported giving and receiving oral sex with all his part-ners but never using condoms for that. For anal sex, most of time he was the receptive partner, or the bottom. Rarely he'd agree to be the insertive partner, or the top, but he didn't consider himself versatile (happy as a top or bottom)—he had his preferences. He usually used condoms with casual partners, but he would often bareback (anal sex without condoms) with a buddy he had part-nered with before, a guy he knew well who tested frequently for STIs and HIV. Most of his partners were one-offs he had met online, guys he didn't know and wouldn't necessarily see again.

With this information in mind, the visit transpired in quick succession. I swabbed his throat and his rectum, sent him to collect a urine sample, and then ordered a blood draw for HIV and syphilis testing. He'd planned on a quiet weekend since he'd be gathering with family early on Easter Sunday; he would wait until he got his test results before he had sex again. In a few days, his results were ready. His tests for gonorrhea and chlamydia (in the throat, rectum, and penis), syphilis, and HIV were all negative.

If conventional wisdom is that STIs result from promiscuity, then why is it that William was afflicted by gonorrhea after having only two partners while an über-active guy like Xavier didn't catch anything?

Multiple factors play into one's risk of contracting an STI: number of partners, sexual orientation, types of sex, race, geography, and socioeconomic status. The interrelatedness of these factors makes it difficult to determine which factor plays the strongest role. For William and Xavier, here's what we know: William was Black; Xavier was Latino. William was straight; Xavier was gay. In the past three months, William had had sex with two people; Xavier had had sex with more than twenty. Neither one of them had consistently used condoms. Yet William had contracted the STI, and Xavier had not.

Was it simply bad luck, as William suggested, or something else altogether?

To arrive at the answer, we would need to study a large group of people having sex just a few blocks away.

Barbells and Blowballs

If you want to see hundreds of sexual acts at once, look no further than the computer of Bob Kohn. Bob is not watching internet porn at the office. As an epidemiologist tracking STIs for the city of San Francisco, what he does at work is quite removed from the act of sex itself. However, sexual imagery certainly abounds in his office suite. One fixture is an eight-foot-tall cardboard penis named Clark that changes outfits throughout the year. Clark is a popular mascot of the city health department at parades and other civic events, and today he is wearing a black vest and harness from a recent stint at the local leather and fetish festival on Folsom Street. Also scattered about are stress-relieving penis-shaped squeeze toys in multicolor flesh tones, remnants of a bygone syphilis-testing campaign. Each penis flashes a winning smile. I surreptitiously grab one, squeeze it, and enjoy a moment of Zen.

I encounter Bob in his usual polo shirt and khakis. With his untamed curls, piercing blue gaze, and deadpan humor, he's a preppy mad scientist meets David Letterman. I share with him my recent tale from clinic and William's "bad luck" theory of STI acquisition. He listens and nods, but I can tell he's not buying it.

He and other sexual network experts assert that the configuration of William's sexual network, or how he connects sexually to others in space and time, is a key factor in explaining why he appears to have misfortune when it comes to contracting STIs.

Bob and I looked at a sample network on the computer. The screen is filled with hundreds of tiny dots connected by lines in a beautiful branching structure. Each dot is a node, or person, with a line, or link, connecting it to at least one other node on the screen. Each link between nodes represents the entire sexual partnership between two people during a specific period of time, whether they had sex one or one thousand times.

One piece of the puzzle behind the tale of William and Xavier can be found within the mass of nodes and links making up this sexual network diagram. To delve a little deeper in our understanding of networks, we first need to learn a bit more about the components that are used to construct them. Even the largest and most complex sexual networks begin with a simple link between two nodes, signifying a single relationship. Here's one example:

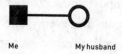

This little barbell represents my entire sex life for the past decade. Sigh. I could go on like this forever and never catch anything. Bob agrees.

While the little barbell of monogamy is generally a low- or no-risk situation, all monogamous relationships are not created equal. The subsequent diagram illustrates a community and its sexual linkages during the prior three months. The two gray squares below represent two different women (Caroline and Alice), who each met a different male partner (represented by a black dot). Imagine also that the man linked to each of them had no other partners once they started their respective relationships.

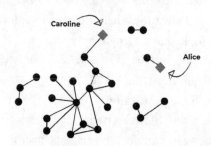

Alice and Caroline each have sex with one man, but those men have very different levels of risk.

While Alice and Caroline were each in mutually monogamous relationships with these partners, it's obvious that their partners' prior sexual networks were starkly different before they entered their current relationships. While Alice's partner was linked only to her in the three months prior, Caroline's partner was previously linked to a sexual network of more than a dozen individuals, both men and women. Even if he only has sex with Caroline now, he could easily harbor STIs from his prior network connections that could be transmitted in his new relationship.

If both Alice and Caroline started their relationships free of STIs or HIV, and assuming they have the same types of sex with exactly the same frequency and similar rates of condom use, it's easy to see how Caroline could be exposed to a higher level of risk than Alice.

Martina Morris, professor of sociology at the University of Washington, utilizes a similar example to illustrate how exposure to an STI or HIV is dependent not only on what you do but what your partners do (and did before they met you), and your partners' partners, and so on, and so on. What happens to you is influenced by the behaviors of people several degrees of separation away from you who don't even know you exist. The next time you find yourself sexually attracted to someone, just imagine a branching network structure protruding from their head. Exercise caution until you can deduce whether that structure resembles a huge oak or a single twig and leaf. (Hint: it will probably take more than one night to figure this out.)

Since I can't catch an STI in a mutually monogamous relationship, I could spice things up by expanding the simple barbell between my husband and me into something more complex. Let's take all the moms and dads in my son's preschool. I would like to know how my network would look if I were able to accomplish that. I'd also like to know how likely it would be for me to catch an STI, such as chlamydia. Since this is happening virtually, there is no risk of actually catching anything, and—bonus—I don't even have to get undressed.

Most of the preschool parents are in stable pairs, or dyads. There have been single moms or dads who are dating, but let's assume they are sleeping with just one person at a time. I do not know of any extramarital affairs, and I do not think there is any swinging going on among the parents. If there is, I haven't been invited to get in on the action.

For this exercise, I'll have sex with each person at least once over the course of a month. I'll stick to one partner per sex act. Although group sex might be more efficient, I'm not that confident in my ability to multitask while naked.

With the prior assumptions of no affairs and no swinging, my hypothetical preschool parent sexual network looks something like this:

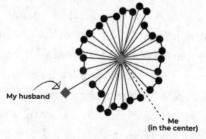

Sexual network of 14 couples, each member of each couple has sex with me as well as with his or her respective life-partner.

That looks a little more promising. In this scenario, network scientists would refer to me as a hub with many nodes. As a single node begins to add connections, the structure starts to resemble a hub with spokes like a bicycle wheel. When the connections get more plentiful, it resembles the puffy seed phase of the dandelion, or the blowball. (Alternative nicknames for this botanical structure—e.g., *priest's crown* or *cankerwort*—seemed equally inappropriate.) In this hypothetical network, I'd have sex with twenty-eight people, plus my husband, if I still had the energy. That should guarantee me at least a touch of chlamydia, shouldn't it?

As evidenced by the prior tale of William and Xavier, there's more to catching STIs than simply having lots of sex partners. John Potterat and Dan Wohlfeiler have been trying to tell the world about this for the past two decades. They're two STI prevention experts from the Colorado Springs department of health and the University of California–San Francisco who study the influence of sexual networks on the spread of STI and HIV epidemics.

Over the years, Potterat and Wohlfeiler determined that the configuration of a sexual network, or *how* people connect sexually in space and time, can predetermine how quickly an infection can spread within a community—and they came up with one of the seminal models for sexual network risk configuration.[4,5] Though it's not terribly complex, it's easier to follow their logic if we know the individuals involved. Let's look at their model by imagining what would happen if Snow White were to be involved in a sexual network with the Seven Dwarves.

In this network, there are eight people connected by nine sexual relationships. Six individuals have two partners each

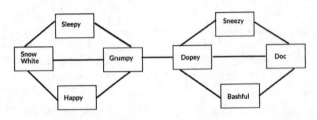

Sexual Network Model 1, Modified from Potterat and Wohlfeiler

(Snow White, Sleepy, Happy, Sneezy, Bashful, and Doc) and two individuals have three partners each (Grumpy and Dopey). We begin with Snow White, who has just been diagnosed with syphilis. We also assume in this network that syphilis is transmitted every time a sex act takes place:

Step 1: Snow White has sex with Sleepy and Happy, causing both to become infected with syphilis.

Step 2: Sleepy and Happy both transmit syphilis to Grumpy. Now half of the network is infected, and half is spared.

Step 3: Grumpy transmits to Dopey.

Step 4: Dopey transmits to Sneezy and Bashful.

Step 5: Sneezy and Bashful have sex with Doc. In five steps from Snow White, everyone has been infected.

But what if the network looked like this?

Count the links and you'll notice there are still eight individuals connected by nine sexual relationships. Again, there are six individuals with two partners each (Sleepy, Happy, Grumpy, Dopey, Sneezy, and Bashful), and two people with three part-

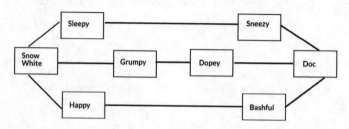

Sexual Network Model 2, Modified from Potterat and Wohlfeiler

ners each (this time, it's Snow White and Doc). But in this scenario, when Snow White gets syphilis, the infection spreads more quickly:

Step 1: Snow White transmits syphilis to Sleepy, Happy, and Grumpy.

Step 2: Sleepy, Happy, and Grumpy transmit syphilis to Sneezy, Dopey, and Bashful.

Step 3: Sneezy, Dopey, or Bashful transmit syphilis to Doc. In three steps from Snow White, everyone has been infected.

In network 1, to disrupt transmission of syphilis from Snow White to Doc, you need only break the sexual link between Grumpy and Dopey. In network 2, you need to break up three relationships (Grumpy-Dopey, Sleepy-Sneezy, Happy-Bashful) to keep the infection from spreading from Snow White to Doc.

Consider William and Xavier again in the contexts of their networks. The network that William revealed to me during our clinic visit is not much different from the simple barbell that I showed earlier, with a single extension for his other girlfriend in San Francisco. Meanwhile, Xavier's sexual network resembles the dandelion blowball shape that we saw in my hypothetical preschool sexual network.

These diagrams show how many people each of them reported sleeping with (or their first-degree connections). We don't know about partnerships that were two, three, or more degrees of separation away from William and Xavier and how they were connected in space and time. As we've already learned, sex

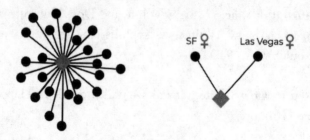

Xavier and William's first-degree sexual connections. Xavier is connected to 25 partners, while William is connected to two.

partners that are several degrees of separation removed can still strongly influence one's risk of infection. What we can deduce, though, is that something about Xavier's behavior plus his sexual network's structure allowed him to stay STI- and HIV-free, despite having unprotected sex acts with multiple people.

Although Xavier had many more partners than William, most of his sexual encounters were one-offs, so each pairing ended before another one began. When we drew out Xavier's sexual history over the past few months, he looked like a dandelion blowball because we were simply drawing out the number of partners he'd had during a given time period. If we look at the order in which he was having sex with his partners, however, his sex life looks more like a string of barbells laid end to end (see the following figure). Network scientists would label him a serial monogamist, even if some of his relationships lasted for just one night.

In the following figures, Ashley (designated by the letter *A*) is a serial monogamist with five sexual partners during a given month. She sleeps with those partners in a non-overlapping fashion, starting with Brian, then Clay. When she has sex with David, who is infected with gonorrhea, she subsequently exposes

Edward and Frank, but not Brian or Clay, who predated her exposure to David.

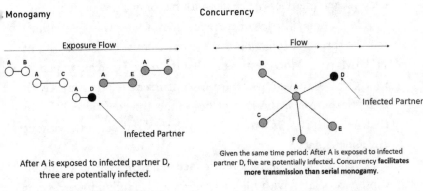

Serial monogamy versus concurrency, Potterat and Wohlfeiler

Next month, her sister Anna has sex with the same five partners but can't bring herself to break it off with any of them, so she goes back and forth between all of them. Anna also gets exposed to David, who is unfortunately *still* infected with an STI, and quickly all five people are potentially exposed. From a disease transmission standpoint, a "clean" breakup touted by psychologists and relationship coaches is crucial to preventing the spread of STIs. The inability to completely end one sexual relationship before starting another may potentially expose many people to an STI in a short period of time.

Network scientists' term for Anna's overlapping sexual relationships is *concurrency*. Concurrency doesn't imply anything about whether one partner has knowledge of the other(s). Concurrency could include simultaneous partners (such as in a ménage à trois), cheating, negotiated non-monogamy, or what we observed in William's case, a don't-ask-don't-tell situation between himself and his two female partners. The more people in a network that practice concurrency, the more rapidly other network members become exposed to sexually transmitted bacteria and viruses, even if each member of the network has only a few partners.

Here's how a little concurrency goes a long way in facilitating the spread of infection throughout a network. In the case of HIV, Martina Morris estimates that if a mere 5 percent of people in a given sexual network are in concurrent relationships with multiple partners, the risk of becoming infected with HIV in that network is *double* that of a similarly sized network without any concurrency.[6]

Even the mere suspicion of concurrency in your relationship can be associated with high risk of an STI. Holly Howard from the California Department of Public Health found that women who said concurrency was "possible" in their relationship were at two- to fivefold greater risk for chlamydia than women who said that it was "unlikely." In fact, women who thought concurrency was "possible" were at the *same* risk as women who said their partners "definitely" had other partners. Bottom line: if you have any suspicion your partner could be sleeping around, you could be at higher risk for an STI.

We know that William was engaged in concurrent relationships with two women, and he suspected that his partners might have engaged in concurrent relationships as well. Adaora Adimora at the University of North Carolina has found higher rates of concurrency among heterosexual Black Americans compared to whites (more on why in chapter 8). This partly explains why bacterial STIs like gonorrhea and chlamydia are five to ten times more common in the Black population than the corresponding white population,[7] despite the fact that, on average, Black women report a similar number of male sex partners to their white counterparts and are more likely to use condoms.[8]

As for Xavier, he had not always used condoms for anal sex. The CDC reports that 75 percent of Latino men who had sex with men had condomless anal sex in 2017.[9] Despite this lack of condom use, Xavier had managed to avoid HIV/STIs—but this lucky streak would likely not endure. Eventually, the high rate of changing sex partners would catch up with him. On average, high rates of concurrency and partner change among men who have sex with men has contributed to overall rates of gonorrhea being eleven to fourteen times higher for this population than among heterosexual men; for syphilis, it's more than one hundred times higher.[10–12]

While exclusive monogamy and serial monogamy are effective strategies for minimizing STI/HIV transmission, they simply don't work for everyone. Some of us struggle with achieving a clean breakup or, like William, would like to maintain partners in more than one locale. Or maybe we simply have an appetite for variety. If you're in one of those situations, there are ways to practice concurrency and remain disease-free. To understand how, let's get a little help from our polyamorous friends.

Inside and Out

Marian is known in her field of environmental science as a brilliant and hyper-organized scientist who thinks quickly on her feet. It would likely shock her colleagues to know that she was once involved in a radically alternative existence, abandoning her career ambitions to immerse herself in the residential headquarters of OneTaste, a polyamorous community currently active in eight U.S. cities and the United Kingdom.

Marian entered OneTaste at a time of explosive growth, as its residential facility expanded from eight people in a single home to a warehouse and community center that housed forty members.

The center of OneTaste's universe was its charismatic entrepreneurial leader, Nicole Daedone. (Her charms are evident in her TEDx talk on the power of orgasm to cure the ills of modern Western womanhood, which has over 1.9 million views.)[13] OneTaste's status as a cult is a subject of debate. But cult or not, what is indisputable is that these kinds of communities provide valuable insights into the practice of sexual concurrency. If the polyamorous like Marian can manage to remain STI-free despite having multiple overlapping partners, perhaps there is something to be learned for the rest of us who find ourselves balancing more than one partner at some point in our lives.

The practice that launched OneTaste is known as *orgasmic meditation* (OM), a fifteen-minute ritual of clitoral stimulation where the recipient (always female) is unclothed from the waist down, while the person conducting the stimulation (male or female) remains fully dressed. For residents of the OneTaste warehouse, twice daily group OM sessions were part of the normal routine. Also consider that forty residents slept in a single common space with no walls, facilitating other sexual activities out in the open at all hours of the day or night. Even if physical boundaries existed during OM, the practicing of OM with ten or more different individuals each week quickly stripped away emotional boundaries that might have been present prior to entering the residence.

From an STI transmission standpoint, OM is essentially a no-risk practice. During Marian's tenure, the community members even started wearing gloves during OM sessions. But for some—Marian included—OM was a low-risk gateway into many other types of sex play. Residents would periodically stage group sex scenes, involving costumes, role-play, oral/vaginal/anal sex, and swapping partners back and forth. This was sexual concurrency as theater.

Given this range of sexual activity outside of OM, an inner circle of community leaders decided that all residents should receive comprehensive STI testing. Although some members had health insurance, the inner circle wanted everyone to have the same baseline testing, so for convenience, most residents went to the nearby public STI clinic en masse.

Copies of everyone's test results were all given to a single record keeper, who ensured people had negative results before they had sex in the community. Marian decided to go to a different laboratory from the others to minimize any chance of running into anyone she knew from outside the community. When she finished testing, she handed her results over to the record keeper. To this day, she still doesn't know to whom her personal health information was disclosed within the community. She continued to test regularly during her residence, which spanned over two years.

Marian estimates that she practiced OM with at least one hundred people during her two-year residence. "During that time, I also had intercourse [vaginal and/or anal] with about 10 men in the community. I know that might sound high, but it's much lower than the number I *could* have slept with." Her recall of her number of oral sex partners was a bit hazy but fell somewhere in between the other two estimates. Ever the scientist, she apologized for the lack of precision in her data. As she explained, she might be blindfolded on occasion during a group scene and couldn't be sure of how many people she had oral or vaginal sex with within the span of an evening.

Despite a sizable number of partners and a very high rate of concurrency within OneTaste, there were no STI or HIV outbreaks during her two-year residence. Several factors likely contributed to this phenomenon. With regular testing and treatment of all members, the chances of having an uncontrolled outbreak would be almost nil, even with very high rates of concurrency.

As Gabriel Rotello describes in *Sexual Ecology*, another factor that affects the spread of STIs/HIV is how much sexual mixing or bridging goes on between members of a group and those from the outside.

In Marian's case, the community was a relatively closed system. Although she described it as having "a semi-permeable membrane," folks generally had sex within the group. Community members did occasionally find partners outside, but the awkwardness involved in explaining the unusual living situation discouraged many from seeking external partnerships. Additionally, she explained, "the longer that you were there, the fewer people outside the community that you actually knew."

Another method used to minimize STI risk in OneTaste and similar polyamorous groups is a practice known as *fluid bonding*. This involves an individual sharing bodily fluid (no condoms or barriers) with only one partner whose STI/HIV status was known and who could be trusted to disclose any changes in that status. In turn, the fluid-bonded partner would agree to the same arrangement. There is no limit to the number of partnerships that could be enjoyed outside the fluid bond, but barriers such as condoms would be used 100 percent of the time with those outside partners.

Marian observed that some members in the community were known to be fluid bonded to one another, but she didn't feel compelled to fluid bond to anyone in particular. Instead, she insisted on condom use for any vaginal or anal sex. She felt comfortable initiating such conversations and had done it so many times that she thought her preferences were common knowledge in the community.

One night, Marian was having sex with another community resident. The sex escalated in intensity, and she was so focused that she forgot to stop and have a conversation about condoms. "I thought [my partner] was reaching for a condom when he was actually reaching for lube." They were having anal sex, so her partner wasn't concerned about pregnancy. Because he was entering her from behind, she also couldn't see what he was doing. She didn't realize that he had entered her without a condom until he climaxed. As he withdrew his penis, "all of a sudden I was like, what is *that* on my back. I was very upset. I blamed myself because I was certain he was reaching for a condom, but it was the one time I hadn't actually had the conversation before the sex began."

Upon realizing that she had just had unprotected anal sex, she abruptly fled. "I went into the community center and just sat by myself and freaked out for a while." She sought testing for STIs and HIV shortly thereafter, testing for HIV multiple times to make sure she didn't miss an infection. Her results were all negative, much like her memory of the experience. Looking back on it now, Marian realized that despite her extensive knowledge and usual insistence about protection, even she was susceptible to the occasional slipup.

Despite the scare, Marian stayed on in the community. As her practice of OM intensified in frequency and length of time, she often felt like she was under the influence of a biochemical buzz. "I was so high. At times I felt like I could operate at 150 percent. I felt incredibly productive." Her elation turned out to be temporary. As she neared her two-year anniversary at One-Taste, she experienced an emotional crash, leaving her depleted and often tearful.

On top of everything else, a financial crisis loomed. For months, all her money was going to the community. One day, her student loan repayment kicked in. Before she realized what was happening, her bank account had a negative balance.

It was time to get out.

Logistically, it turned out to be relatively straightforward. "I called my family, got a loan, called up my boss, and went back to my prior job." Emotionally, it was much messier. Marian still had relationships with many coresidents who were roommates and also ex-lovers and friends from whom she found it difficult to disentangle.

Her final push out the door happened when someone she was deeply attracted to came back into her life, an old flame from college who had moved back to San Francisco. "I realized that if I wanted to be with this other woman, I needed to get out. I needed a hook out of the community, and she was it."

While Marian has largely put her polyamorous life behind her, occasionally she encounters a ghost from the past. "Sometimes I run into someone who looks familiar and I just can't remember, do I know them from work, or did I have sex with them? It's uncomfortable, but it's becoming less frequent over time."

Despite some negative memories of her experience at One-Taste, Marian doesn't discourage other people who want to be in concurrent or polyamorous relationships with multiple partners. What she does advise, though, is taking calculated risks. For someone who is active with partners concurrently, this means fluid bonding with just one person (if at all) and using protection with the rest. It also helps to test frequently and share results with partners. Making intentions known about safer sex *before* becoming aroused, with every single partner, is ideal. But Marian learned to forgive herself and others who couldn't manage to have that conversation every time. "Even the most dedicated and educated of us aren't perfect. I'm just lucky."

Sleeping in the Core

Meanwhile, I'm still sitting in the center of my preschool sexual network, futilely trying to catch chlamydia. Despite sleeping with more partners at the same time, the average age of individuals in my network is over thirty-five years old—the prevalence of chlamydia in this age group is less than 1 percent (0.7 percent to be exact).[14,15] The *prevalence* of an infection, or how common it is within a population, also strongly determines how likely it is that one might become infected if one has sex within that population. In my network, where exclusive monogamy usually prevails, it's likely to be close to zero. I might have a decent chance of catching a viral STI such as herpes or HPV (chapters 1 and 4), but if we are looking at the major bacterial players, such as chlamydia, gonorrhea, or syphilis, this network would not provide an easy way to get infected.

In other sexual networks with a higher prevalence of STIs, infections don't distribute evenly or spread at the same rate among all the individuals in the network. A few key individuals serve as *core transmitters*, who are responsible for facilitating the rapid spread of an infection throughout the network structure. HIV researcher Robert Grant prefers not to label these core transmitter types as "high-risk" individuals but refers to them instead as "affectionate and popular," as if they're the star quarterback or homecoming queen of intercourse. It's these few affectionate and popular core transmitters, or *core group members*, that are also responsible for sustaining the STI epidemic once it has begun.

While having multiple partners often characterizes members of the core, author Gabriel Rotello (*Sexual Ecology*) observed that "perhaps equally importantly, those partners also have significantly higher numbers of partners *within the core*, creating a kind of biological feedback loop that is primed to magnify disease."[16] One of the best-known core groups was centered on French Canadian flight attendant Gaétan Dugas, widely labeled as "Patient Zero" of the AIDS epidemic. Evolutionary biologist Michael Worobey's analysis of HIV's genetic diversity would reveal, in fact, that Dugas did not introduce HIV into the United States,[17] but he and his partners were certainly part of a highly sexually active core group who fueled the initial epidemic.

Let's think back to the tale of William and Xavier. William's repeated bouts of gonorrhea not only exemplify the effects of concurrency but imply that he, his sex partners, and his partners' partners were also part of this feedback loop of core transmitters referred to by Rotello, leading to repeated infections in a short span of time. On the other hand, Xavier's partners were certainly plentiful, but it's possible that most weren't members of a high-risk core group. Combined with his pattern of serial monogamy (mostly one-night stands) and perhaps just a little luck, he managed to stay out of a higher-risk network.

If we are trying to reduce high rates of STIs within a population, reaching this core group of transmitters is key; we'll never make a dent by capturing random infected people here and there. The way to disrupt the core would be either: a) convincing core transmitters to change their ways or, at the very least, test themselves frequently, or b) getting others to avoid having unprotected sex with core transmitters.

So how can one identify a core transmitter? According to Charles Fann, a former syphilis investigator in San Francisco, it's not that easy. He knows who they are of course; they are the folks whose charts cross his desk several times in a year because of repeated infections. But it's not like you can pick them out of crowd; they don't always have good looks or a sparkling personality. And core transmitters aren't necessarily single. According to Fann, a third of core transmitters in San Francisco are in a primary relationship (including marriage), but as Savage Love columnist Dan Savage describes, these relationships are "monogamish," where the relationship is open on at least one end.

One defining characteristic that Fann described of core transmitters was a sense of allure, a self-confidence and comfort with themselves that they pass on to their partners. He says you can't necessarily put your finger on it but that you can feel it when you're in the room with them. His clients who have had sex with a core transmitter report doing things with them that they wouldn't ordinarily do. Sometimes drugs or alcohol complicate that picture, but more often, people take risks because the core transmitter made them feel so good about themselves in a non-contrived way. All the usual cautionary thoughts in one's forebrain—"I'm not looking for a random hookup tonight" or "I should use a condom"—fly right out the window.

If the allure of core transmitters is so irresistible, then should we just throw in the towel and submit to the power of the core? Fann doesn't think so. He still believes in the role of counseling to strengthen his clients' resolve for practicing sex more safely— so that later, if a client meets someone they find really alluring, they might have enough agency to negotiate what they want before arousal pushes them to the point of no return.

Without a reliable way to proactively identify core transmitters, we are left with identifying them after the fact. This is the role of the health department, who tries to interview patients with certain STIs such as syphilis and trace all their sexual partners within the past three months (more in chapter 6). As each sexual partner is tested, interviewed, and asked about their partners, connections are revealed, and a network diagram is gradually constructed like an image in a puzzle, one piece at a time.

In the middle of a disease outbreak, the health department

might even add in a technique called *clustering*, asking you to name sex partners *plus* other people you associate with (nonsexually), because you and your associates may have sex partners in common without realizing it. Imagine sorting through all this information manually, eventually creating network diagrams of the complex sexual connections between infected and uninfected people. Although it's a tried-and-true way to identify core transmitters, it's also arduous and time-consuming.

It took fifteen years, but Bob Kohn figured out how to automate this process. He programmed his computer to sort through hundreds of people who had been named as sex partners of syphilis cases and placed them into sexual networks. It was time well spent; his program allowed him to characterize one of the largest sexual networks described to date, referred to by Kohn and his colleagues as *Mega-Network*.[18]

In Mega-Network, multiple dandelion blowball structures abound, with the hub of each blowball containing a core transmitter who had syphilis. As Mega-Network began to form throughout 2013, had the syphilis infections stayed trapped in small components on the periphery, it would not have been transmitted efficiently throughout the network. Instead, it was able to enter the center of the network, where the links between people are dense and plentiful. This happened due to other core transmitters who served as bridges, connecting components of the network to each other. A bridge's connections need not be plentiful but are just as crucial to creating the network structure as those hubs with many partners (see following figure).

A decade before the birth of Mega-Network, network scientist Albert-László Barabási noted in his book *Linked* that "obtaining a map of the sex web, which links people via sexual relationships, is simply impossible."[19] He couldn't believe that

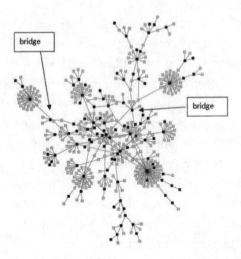

Mega-Network of 435 interconnected sexual relationships in a one-year time period, San Francisco, 2013. Black squares represent individuals who had syphilis alone or a combination of HIV/syphilis. Gray squares represent uninfected (or not tested) sexual contacts. From Robert Kohn, with permission.[18]

people would be willing to give up the names of everybody with whom they had been sexually involved, knowing that someone else would then try to map out all their sexual contacts. Barabási might be shocked to know that when Kohn applied his algorithm all the way back through 1997, Mega-Network grew to over 6,500 connected members by triangulating names of partners disclosed from patients infected with syphilis. Although Mega-Network is likely missing some people and their partnerships, it is a remarkable example of the power of network analyses to study disease transmission.

Fann and Kohn have put their heads together to see how Mega-Network can augment Fann's disease investigation activities. Fann added some network-speak to his counseling: "This is a high-risk network, and you are in it. I'm not telling you to stay in or get out. If you stay in it, here is how we can support you." He offers a menu of prevention options, including HIV pre-exposure prophylaxis (see chapter 9), postexposure prophylaxis, partner notification, free condoms, linkage to medical care, and free STI/HIV testing. He's made peace with his ability to change the behavior of core transmitters and people who have sex with them.

"Some people are getting STIs as a by-product of the type of life they are living," he said. "As long as they are coming in at an early stage of infection and routinely screening [every three months], maybe that is the best that you can do. Ultimately, you can't make anyone protect themselves."

Kohn also plans to use network maps to search beyond a patient's recent sex partners to include people who are separated by one degree (those who slept with a patient's sex partners but didn't sleep with the patient). He and Fann plan to see if they can find these sexual "associates" and whether any of them will respond to outreach. "We can't tell them, 'We know you've been exposed,'" Kohn said. "We'll tell them something like, 'We've identified syphilis in your sexual network' and see if we can find any disease among these people." It's the equivalent of saying, "I'm not sure syphilis is at your doorstep, but it's going around the neighborhood."

Cynics ask Kohn whether any of this work can stop the current epidemic increases in STIs. He thinks maybe the right system of mapping networks could help. He is determined to find such a system, even if it takes him fifteen more years. There are new computing tools to help him in his quest, but we don't yet know whether using networks to predict future connections will be a game changer in STI control.

Until Kohn and his colleagues figure it out, we must still rely on good old-fashioned disease investigation: asking patients with STIs to name their partners and trying to notify them they've been exposed. While it sounds straightforward, public health disease investigators like Charles Fann know otherwise; getting someone to fess up about their sexual escapades is no mean feat.

6

Knock, Knock,
It's the Sex Detectives

Tales of Public Health's Heroic Contact Tracers

I'd Tell You, but Then I'd Have to Kill You

John Potterat heard his phone ring and grabbed the handset, the cord stretching as he brought the receiver to his ear. His brunette locks were styled into a mullet with long bangs, his zigzag patterned shirt at odds with his diagonally striped tie. Next to his phone was a zebra-print mug and a large ashtray. On the wall behind him, he'd affixed a bright yellow bumper sticker that advised, SUPPORT YOUR LOCAL HOOKER.[1]

The year was 1974, and Potterat was staffing the only desk in the lone office of the Colorado Springs health department's venereal disease (VD) program. At the time, the program's budget could support two staff, Potterat and his colleague Christopher Pratts. While others referred to them as sex detectives, their official title was a bit more formal: disease intervention specialist (DIS).

Their job was to interview patients from the city's VD clinic who'd been diagnosed with gonorrhea or syphilis, also known as *index patients.* During the interview, they'd elicit names, phone numbers, and addresses of the patient's sexual partners, referred to in the business as *contacts,* from the prior two to three months. Potterat and Pratts would then go out into the field to track down as many of the sexual contacts as they could, a process known as *contact tracing.*

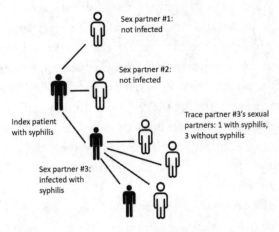

Contact Tracing: an index patient with syphilis names three sexual contacts in the last three months. Disease Intervention Specialists (DIS) find the contacts and test them for syphilis. An infected partner is found and interviewed for their sexual contacts. The process continues until there are no more contacts left to test (a patient or contact refuses to participate, all contacts test negative, or no more contacts can be found).

The practice of contact tracing was born in 1937, when U.S. surgeon general Thomas Parran first proposed its use for the control of syphilis. Over the years, health departments also used it as a tool to control outbreaks of tuberculosis, hepatitis, and most recently, novel coronavirus (COVID-19). Regardless of the infection, the principles were the same. The role of the DIS was to figure out which contacts were upstream and might have infected the original patient and which contacts were downstream, to whom the patient might have transmitted the infection. By testing contacts and establishing when they had sex with the index patient, the DIS could tease out how contacts were related to the patient and sometimes to each other.

Once they located a sexual contact, Potterat and Pratts would try to convince them to come to the health department's clinic for testing and treatment. They would also treat any contacts who came in before their test results came back; this would address any incubating infections before they became symptomatic. If one of the partners tested positive for an STI, they'd interview that person as well, tracing the chain of infection until all potentially infected partners had been treated. The faster they could find and treat contacts, the more quickly they could interrupt transmission of the infection, like putting out a flame before it becomes an inferno.

If a DIS approached you and said you'd been exposed to syphilis, your logical first question might be: *Who* gave this to me? But like all disease intervention specialists, Potterat and Pratts would never disclose any information that could reveal the index patient's identity. Even if you'd only had one partner in the past two to three months ("I just know it was him"), Potterat and Pratts could never confirm your suspicions.

With just two staff, everything in the Colorado Springs VD program would fall to them, running off the engine of their mutual enthusiasm. In addition to patient interviews and contact tracing, they would count cases, write grants and reports, and handle the secretarial work. When Pratts was out in the field searching for someone's sex partners, Potterat had to stay by the phone, responding to calls from patients inquiring about their test results or fielding questions from the public, such as: "I have a sore on my penis, what should I do?"

One day, the caller was a VD clinic patient inquiring about her gonorrhea test results. After delivering the woman's positive result, Potterat flipped to the section of a ledger that had handwritten vertical tally marks, each one representing a case of syphilis or gonorrhea. This was their surveillance data, the way he and Pratts counted cases reported to the health department that year. Potterat put another tally mark in the ledger, adding another case of gonorrhea to the year's total.

The patient promised Potterat that she'd come to the VD clinic that afternoon for treatment. His office was adjacent to the clinic's waiting room, so Potterat was able to pop by for an interview while she was there.

Sometimes these interviews were simple. The index patients knew their sexual contact(s), knew exactly where to find them, and were willing to give that information to the health department. Better yet, some patients would take on the task of disclosing the diagnosis themselves: "Let me tell that cheating [insert expletive] myself so I can hit him upside the head before I send him to your clinic."

Usually, though, it wasn't that easy. Finding a contact might involve searching all over town—not only visiting people's homes but also bars and burger joints, street corners, drug houses. A DIS had to be prepared to walk into any situation with complete strangers and keep their cool as they tried to locate a patient's sexual contacts.

Often, they were missing key information. An index patient might say, "I don't know the name of the guy I slept with, but he's about five foot ten, with dark brown hair and brown eyes. He was from California. He works in a pizza place, somewhere on

Platte Avenue." Potterat would drive over to that street looking for pizza restaurants, go inside and check out the employees, look for someone meeting the description, subtly pull them aside, and ask, "Hey, are you from California?" to see if he could find the right person.

If Potterat was certain he'd identified someone, but he couldn't speak to them privately, he'd hand them an unmarked envelope with his contact information and a message inside, stating that he needed to talk about a serious health issue. Discretion was key; the note would not mention the words *sex* or *VD*, nor would he speak those words aloud in public. He'd never flash a badge to intimidate people into talking. He was not the sex police—the process of interviewing and contact tracing was voluntary.

If you knew that a stranger with a mullet would appear at your workplace to deliver news of an STI exposure, you might try to avoid him. This is where the element of surprise worked in Potterat's favor: people didn't know he was coming until he was already there. If it was safe to talk, he'd introduce himself and give his standard speech stating his credentials and ensuring confidentiality, then he'd gently break the news: "You've been exposed to syphilis. I'd like to bring you into our clinic to be tested and treated." Sometimes people were shocked and angry; occasionally, he had a door slammed in his face. But Potterat was as disarming as he was persistent; he could get almost anyone to talk eventually.

One key to his success was empathy. People sensed it and opened up to him, not only about sex but also about the other ills affecting their lives: abuse, addiction, money troubles. When he spoke, he spoke plainly, without any jargon. "I would talk the same way to a seven-year-old child as I would to the pope." Regardless of a person's occupation, race, or class, "what I saw in front of me was a person. Somehow, they knew."

He also knew which levers to pull for information. If there was a cluster of gonorrhea cases involving local prostitutes, Potterat knew he'd first have to approach the pimps. On one occasion, a pimp wanted to size him up before he would talk, challenging, "Hey, white boy, can you shoot pool?" (He could, but poorly.) The pimp put a twenty-dollar bill down on the table and said, "I can beat your ass." So Potterat put his twenty down there "knowing I was going to lose it, but I was going to play."

Potterat lost the money but gained the pimp's respect. Subsequently, the pimp provided names and numbers of the women who worked for him, and Potterat got them to come to the VD clinic for testing and treatment. Word got out that Potterat was okay, and soon the prostitutes and pimps were in his confidence. He knew that sex work didn't adhere to normal business hours, so he often gave out his home number (these were the pre–cell phone years). His kids grew up fielding calls from prostitutes and pimps at night and on weekends.

Another group at high risk for STIs were the local motorcycle gangs. In biker gangs, sexual concurrency was common; the bikers and their girlfriends often swapped partners, allowing STIs to spread quickly through their sexual network.

Over the years, Potterat became friendly with the bikers who were "frequent flyers" at the VD clinic. One day, Mike, a leader of a local gang, approached the clinic's registration desk in a state, demanding to be seen immediately. "I don't want to wait with these 'dirty' people," he said. "I'm very busy."

The receptionist informed Helen Zimmerman, the clinic's diminutive nurse practitioner, who, Potterat lovingly recalled, "took shit from nobody." She'd warn bikers like Mike packing knives to put them away before they "cut the thing you love most." Zimmerman got right into Mike's face, scowling up at him. "Now you listen here," she started in on him, scolding him to be more considerate. Mike backed down immediately. Perhaps it was her tone and the look in her eye. "Motorcycle gangsters have mothers too," she reflected.[2]

After Zimmerman examined him, Mike walked next door to Potterat's office to say hello and was smiling with relief. Zimmerman had diagnosed him with urethritis (inflammation of the urethra) and given him some antibiotics, reassuring him that it was nothing serious.

Two weeks later, Potterat ran into Mike after work at a rock 'n' roll club in downtown Colorado Springs. Mike slapped him on the back and said, "Man, I feel so much better. I can pee without having it hurt." To add credence to his claim, he asked, "Do you want to see?" Potterat laughed and politely declined, adding, "If I saw it, I'd feel jealous. You're probably twice as big as I am."

Potterat saw this as the heart of his work, building relationships with key members of the community. Connections like Mike became trusted allies, informing him of developments in town: new drugs on the scene, new gangs coming in from outside the state, all information that might affect local STI transmission patterns. Potterat's connections were often highly sexually active and socially connected. In a sexual network diagram, they would be like hubs of a wheel with many spokes protruding from the center.

In 1977, Mike contracted one of the first cases of antibiotic-resistant gonorrhea in the state of Colorado. This strain of gonorrhea originated from East Asia and was highly resistant to penicillin. Mike was alarmed. He'd heard stories during the Vietnam War era about incurable VD; there had been rumors that soldiers were sent away to a remote island if they contracted it. Mike worried the government would quarantine him somewhere until they could figure out a cure.

Potterat had served a tour of combat duty in Vietnam and reassured him that such an island didn't exist. The sex police were not coming to take him away. "Look, drop your pants, we'll give you a shot in the butt, and you'll be fine." Luckily, they had another antibiotic, spectinomycin, which was effective against this strain of gonorrhea. Luckily as well, Mike understood what an antibiotic-resistant STI might do to his sexual network. He named all his sex partners and insisted on escorting each of them personally to the clinic to be examined, treated, and then interviewed by Potterat or Pratts.

At Mike's urging, his sex partners disclosed their sexual contacts, which included other members of Mike's gang. Mike then insisted that his gang members submit to exams and interviews and give up their partners' names, plus others outside the gang that had slept with the same women. Potterat and Pratts spent three weeks interviewing, testing, and treating Mike's sexual partners and his partners' partners, until the chain of infection was finally broken.

Mike was so grateful that he wanted to pay Potterat back somehow. He pledged that if Potterat ever needed anyone killed, he would take care of it personally. Potterat chuckled, dismissing the comment as biker braggadocio. When Mike left his office that day, Potterat didn't realize that it would be the last time he would ever see him.

Six years later, Potterat sat in the break room at the health department leafing through *The Pueblo Chieftain* newspaper, left behind by a nurse who commuted between Pueblo and their clinic in Colorado Springs. The cover story told of motorcycle gang members headed to prison for serving as contracted hitmen for the Mafia. Included in the list of names was Mike's.

Potterat prided himself on possessing steely nerves. "When I was in Vietnam, I was a good soldier, in the sense that I had no breakdowns. I was a solid person psychiatrically." But something dawned on him: "My God, Mike once told me that if I needed somebody killed, he'd take care of it." This was more jarring to him than anything he'd experienced during the war; he began to shake uncontrollably, and it took him forty-five minutes to calm himself down.

As rattling as that incident was, it did nothing to dampen his enthusiasm for his work. Potterat spent eighteen more years in the field, eventually becoming the head of the Colorado Springs STI/HIV program, leading groundbreaking studies on sexual networks, contact tracing, gonorrhea transmission, and prostitution. He gave up on the mullet and stopped doing fieldwork, but his insatiable curiosity about people never wavered.

Potterat offered advice to future generations of disease intervention specialists training under him; to understand the larger social context of STI transmission, a DIS must become a closet anthropologist and ethnographer. This meant putting their boots on the ground in places and neighborhoods where transmission was actually occurring. "See and be seen," Potterat would tell his staff.

Throughout the late '70s and early '80s, dozens of his colleagues lived by this mantra, spending hours in the streets, embedding themselves in the communities at greatest risk of STIs. It was a carefree time, pre-HIV/AIDS, when syphilis was the worst diagnosis they ever had to deliver. Contact tracing was arduous work, but a DIS could still have fun "being seen" out on the town while getting to know index patients and their sexual contacts. What Potterat wouldn't find out until years later was how some of his colleagues were taking his advice of "see and be seen" to another level.

In the Swing of Things

I ... I just don't know what to do."

One Monday morning in 1983, DIS Nancy Spencer listened as her colleague lamented over what had happened at the VD clinic in Denver on Friday night. Each of the DIS took turns being on call during the clinic's evening hours in case a patient with syphilis happened to walk through the door. The patient would be treated with a painful shot in the butt of penicillin, and the DIS would seize this opportune moment to initiate a contact tracing interview.

Last Friday, Spencer's colleague had attempted to interview a man named Rick, who'd been diagnosed with syphilis. Things had not gone according to plan. He didn't have information about his sexual contacts because he'd had sex with them in a private swingers' club, held in a home in the suburbs of Denver. The colleague was flummoxed but got Rick's phone number and told him that they would get back in touch.

Spencer was intrigued, and given how many people could be involved, she wanted to figure out what was going on. "Give me a whack at him," she told her colleague.

Spencer called Rick up and asked him to come back for a second interview. They sat down together and spoke at length. Rick was shaken up by the syphilis diagnosis and wanted to find a way to notify his anonymous partners, but Spencer didn't get any further than her colleague had. "He didn't know who he was having sex with. He really didn't know."

Then Spencer had an idea: Why not have Rick call some of his swinger friends and see if they might know some of his partners' identities? He immediately called up two friends, and despite their shock at the news, they promised to be tested and help notify others in the club. Unfortunately, neither of them knew the names of Rick's partners.

Another solution hit Spencer: "Well, can we come out *there*?" Rick was game. He called the group's manager and convinced them to allow the health department's contact tracers to offer syphilis testing at the swingers' club the next Saturday night, the busiest night of the weekend.

Spencer let her supervisor know about the case. Her supervisor was supportive, but he thought it would be safer if she had company. Her colleague Ana was interested, and they wanted to add a man to the team as well. Spencer knew her gay colleagues would not be interested in working a case at a straight swingers' club, so she called up Robert, a fellow DIS based in Pueblo. Over the years, he'd earned the nickname Kinky Bob, because he relished "weird, way-out situations," like the one they were currently facing. Kinky Bob's response was immediate: "Oh, this is so out there. I'm coming up to work this case with you."

On Saturday night, the three pulled up to a sprawling suburban ranch house in Denver and rang the doorbell. Rick had clearly paved the way for their arrival; a few towel-clad swingers cordially greeted them as they walked by. One of the swingers took them on a tour of the house so they could see how things were set up. As they passed other members of the club, their host would say, "Here are the people from the health department." People nodded and smiled, promising to come by and say hi later.

There was nothing kinky about the setting or the décor. Each room had low lighting, a double bed neatly made with fresh linens, and simple furniture. The only distinguishing feature of the house was a room with an indoor hot tub. After their tour, their host brought them to a brightly lit kitchen and invited them to make themselves at home.

Spencer had brought supplies to draw blood for syphilis testing and a log sheet to record people's information. Along with Ana and Kinky Bob, they set up a testing area at the kitchen table. One by one, the swingers began to drop by. Spencer and her colleagues each took turns greeting them while the others drew blood and logged their information. They handed out cards with the health department's phone number so the swingers could call the next week for their test results.

Everyone was warm and welcoming, glad the health department was trying to help. Still, Spencer felt awkward and slightly out of place. She looked down at her street clothes, then at all the people milling around just wearing towels. She piped up, "I feel kind of silly with clothes on." She stood up, walked to the nearest bathroom, took off her clothes, wrapped herself in a towel, and came back to the kitchen. Ana and Kinky Bob followed her lead, and once in towels, they all sat down and resumed drawing blood. Now on equal footing with their hosts, they felt more relaxed. Demand was brisk that evening, and the three of them continued their work until well after midnight.

As they were finishing, Rick came by. "You're done? Why don't you guys come and join us in the hot tub." As they walked through the common areas, they glanced around to see if people were cruising for sex or having it out in the open, but saw nothing. The agreements to have sex and the acts themselves were happening behind closed doors.

The three of them walked over to the hot tub room with Rick and a few members of the club. They shrugged at each other as if to say, "What the heck, we've already come this far." They dropped their towels and hopped into the hot tub, soaking with the swingers before heading home.

The next day, Kinky Bob went back to Pueblo, his penchant

for weird situations fully satisfied. Spencer came to the office full of stories on Monday morning, and her supervisor and colleagues were supportive of what they had done. Later, Spencer would reflect on the episode as emblematic of a certain time and place in their lives: "The culture was different because that was the early '80s, before HIV/AIDS, before awareness of sexual harassment. People did crazier things. Today, I wouldn't do that. I wouldn't even get in a hot tub with my friends now naked, much less coworkers and strangers."

There was still a mystery to solve: Which of the swingers had infected Rick with syphilis? Later that week, Spencer received a call from the laboratory with the swingers' test results. They were all negative. She wondered what could have happened—if none of the swingers had syphilis, then where had Rick contracted it? Were there other partners that he hadn't told them about? And how had he not passed it on?

"We never found out where it came from," she said. Even with the cooperation of Rick and his contacts, they had hit a dead end. A DIS always investigated contacts as thoroughly as he or she could, to deduce the timing and directionality of an infection. But dead ends were common in DIS work, and sometimes they were created by index patients themselves: they'd give a partial list of their contacts, lie about the names, or give fake addresses. Even though these interviews were completely voluntary, people sometimes just told her what they thought she wanted to hear rather than refusing to talk.

Certain types of interviews were particularly difficult because they brought up issues of infidelity in "monogamous" relationships. Such was often the case with women who had gonorrhea—these interviews never seemed to go anywhere or yield much. No matter how hard Spencer tried, these women typically did not want to give up the names of their partners. And when they did, the partners rarely agreed to come in for testing and treatment. Occasionally, a patient's partner was furious when they heard they'd been exposed to an STI and got hostile with Spencer. At one point, she and a colleague were physically assaulted by a patient's angry partner.

Sometimes the stress of her job would burn her out: "I would go into vacation thinking, I hate these things [interviews]—awful. Then I came back refreshed, I did some interviews, and I thought, this is okay. Somehow, I got this second wind and I could keep going."

She sustained herself by focusing on what she could do well, which was getting people tested and treated for STIs. As for her clients' other life circumstances, she knew they were often struggling with relationship troubles, incarceration, addiction. She tried to let go of her frustration over what she couldn't control. "I felt like I was there to do something and I could only do that one thing and I would be darn good at it. I can't do the rest of it. I can't fix their lives."

Spencer's job frustrations would soon be dwarfed by the enormous tragedy that would overwhelm the public health world. In 1981, she and John Potterat both read the Centers for Disease Control and Prevention's (CDC) early reports of severe immune deficiency in young, previously healthy gay men, now well known as the first cases of AIDS. At the time, they weren't sure what to make of it. Potterat wondered whether a batch of contaminated street drugs was to blame. They didn't realize that epidemiologists would ultimately uncover a sexually transmitted virus as the root cause of these deaths or that this virus would threaten to kill the practice of contact tracing as they knew it.

On Death's Door

By 1983, case reports of AIDS patients began to trickle in to Nancy Spencer at the health department in Colorado. Sometimes the patients were so ill they might die before she could interview them. The investigations of these early cases might have ended there, but each was a crucial source of information on potential risk factors for AIDS. For the first time in her career, Spencer had to conduct postmortem interviews of an index patient's family and friends. She had to maneuver carefully to protect the patient's confidentiality, however, as they might not have revealed the true cause of their illness and death to others.

For those who lived with AIDS long enough to be interviewed, the sessions were heart-wrenching and draining. Spencer and her colleagues' caseloads began to climb. Their gay colleagues began to fall ill. The pace caused some of them to burn out and quit, as the situation became ever grimmer. In *Life* magazine's piece, "The AIDS Tracers," her colleague Kinky Bob told journalists he'd advertise the toll of the AIDS epidemic on public health workers on a T-shirt: I DIED FROM AIDS, AND I DIDN'T EVEN HAVE IT.[3]

Before 1985, the consensus was that there was no point in telling sex partners of AIDS patients that they'd been exposed to a deadly contagious disease. This was the sentiment among public health workers at a time when there was no screening test available for HIV, nor any treatment for those who were ill. A DIS would still interview an AIDS patient to ask about their sexual practices and drug use but stopped asking patients for names of their sexual contacts. This spared their contacts the agony of hearing they'd been exposed to a fatal illness, with no way of finding out if they were actually infected.

After HIV antibody testing became available in 1985, a few states like Colorado resumed contact tracing for HIV/AIDS, but their efforts were met with hostility. Gay rights advocates feared that the practice of contact tracing would allow government agencies to serve as Big Brother, compiling lists of gay men that might be used to target them later. Advocates fought fiercely to protect the privacy of gay and bisexual men, even from the good intentions of public health departments.

Given the climate toward homosexuality in the United States, these fears were certainly justified. In 1986, the Supreme Court ruled in *Bowers v. Hardwick* that anti-sodomy laws were constitutional, and twenty-four states had such laws on the books, which would stay in place until 2003. Some states included both anal and oral sex under the definition of sodomy. (If oral sex is sodomy, then under the legal definition, 85 percent of Americans are sodomites.)[4]

With the introduction of an antiretroviral treatment (which inhibited HIV replication) with AZT in 1987, there was hope that it would prolong the lives of patients with HIV/AIDS, and a feeling that it was worth notifying their partners that they'd been exposed. However, health departments continued to be leery of contact tracing. In 1988, all fifty states had the ability to perform contact tracing for HIV/AIDS, but only twenty-two states actually conducted the practice. Most of these states were in areas with low rates of HIV infection.[5]

Given the labor-intensive nature of contact tracing and the years an HIV infection could exist before patients developed symptoms, researchers such as George Rutherford from the University of California–San Francisco feared for the sustainability of the practice: "If one were to attempt contact tracing for all sexual and needle-sharing contacts of all persons with HIV infection in cities like New York or San Francisco, the costs would be enormous and would likely eviscerate all other prevention and education activities."[6]

Suddenly, contact tracing, a mainstay of public health disease control for forty years, had become, according to John Potterat, "the sickly man of public health." Potterat, one of the nation's staunchest supporters of the practice, wrote editorials urging the scientific and public health communities not to abandon the practice: "Contact Tracing's Price Is Not Its Value," "DIS as Corps Not Corpse," "Partner Notification for HIV: Running Out of Excuses."[7-9]

He conceded that contact tracing could benefit from reputation management, acknowledging that the term smacked of snooping by a government agency. He reluctantly accepted the name change proposed by others in the field, rebranding it to *partner notification* or *partner services*, more customer-friendly terms reflecting individual-level benefit from informing partners about their HIV exposure. Ultimately, though, he felt that *partner notification* was a "unidimensional term for a multidimensional activity."

To Potterat, the heart and soul of this public health intervention was not just notifying partners of an STI or HIV exposure—it was the rich information about social and sexual networks that could be gleaned by tracing and mapping out people's contacts. Mapping out these connections provided DISs and epidemiologists a big-picture view of the STI/HIV risk in multiple sexual networks—an ability to see the forest for the trees.

With the advent of highly active drug cocktails for HIV in the late 1990s, HIV/AIDS transformed from a universally fatal illness to a chronic disease, one that could be managed and controlled. The stigma around HIV began to fade, but there was still trepidation around contact tracing/partner notification. In a national survey of health departments in 1999, Matthew Golden at the University of Washington estimated that only a third of patients with new HIV diagnoses were interviewed to elicit names of partners, compared to 89 percent of patients with syphilis who received similar services.[10,11]

And by the time health departments began warming up to the practice of partner notification, the game had already changed. In 1999, inside a chat room in the city of San Francisco, twenty-three men transformed the practice of contact tracing and partner notification forever—and they didn't even realize that they had done it.

The Sisyphus of Syphilis

In the ancient Greek myth of Sisyphus, the gods punish the cunning and deceitful king of Ephrya by forcing him to roll a giant boulder up a hill, then watch it roll back again, an action he must repeat for eternity. Jeff Klausner didn't actually spend his workdays in a loincloth pushing boulders uphill, but his friends wondered if he identified with the condemned king. Klausner was the newly minted director of STD prevention for the San Francisco Department of Public Health, a job aspirationally referred to by the misnomer *STD controller*. There was widespread doubt that he'd be able to control STDs in a city known for unfettered sexual expression, but he was game to give it a try.

Klausner loved starting his workweek by seeing patients at the city's STD clinic. One Monday morning in the spring of 1999, he was interviewing a new patient named Bill, an HIV-positive gay man in his late forties.

"How many new sex partners have you had in the past six months?"

"Fourteen."

"Okay, how many new sex partners have you had in the past two months?"

"Fourteen."

"So what happened two months ago?"

"I got online."

In the late 1990s, the internet was ruled by America Online (AOL), the nation's largest service provider. Bill had signed up for AOL's monthly subscription service and quickly discovered an online chat room called SFM4M (San Francisco Man 4 Man). Men who were looking to hook up with other men could enter the chat room and initiate conversations with whoever happened to be there; at any given time, a chat room could hold up to twenty-three members. Using AOL's Instant Messaging (IM) feature, men could swap information: age, hair and eye color, height, weight, chest and biceps circumference, penis size, and preferred sexual activities. Then they could agree upon a meeting place where they could have sex.

Until Bill discovered AOL, he had been having trouble finding sex partners. He felt that being bald and a little overweight didn't help matters. Then there was the issue of disclosing his HIV status, which was a deal breaker for some potential partners. On AOL, Bill could enter the chat room, meet a partner in a few minutes, share his statistics and HIV status, then message the guy his phone number. Then they would briefly talk over the phone before arranging a meetup.

Klausner described how Bill would arrange a hookup. "If he liked their voice, he'd invite people to come on over to his house. Then he'd sit and peek out the window. If he liked what they looked like, he would answer the door. If he didn't like what they looked like, he would pretend he wasn't home."

Once Bill invited his guest into his home, sometimes they would chat for a little bit, but other times, they got right down to business. "Sometimes he would get in his bed and turn the lights off and there would be no verbal communication whatsoever. But he said it was a very safe and efficient way to have convenient sex hookups without having to spend money, without having to go to a bar or a club, and he thought it was great."

According to Bill, he could initiate and realize a hookup in less than fifteen minutes. Bill later related to Klausner, "It's easier to get a new partner in San Francisco than it is to get a pizza delivered." Bill seemed pleased with his recent discovery. Klausner, on the other hand, was filled with a sense of foreboding. He knew he was witnessing a shift in the sexual landscape—he worried there would be no looking back.

In 1999, the CDC launched a syphilis elimination campaign, the fourth of its kind in the history of the United States. At the time, syphilis elimination seemed like an attainable goal; cases were at all-time lows all over the country. Congress had appropriated $30 million in support of the CDC's campaign. Half of all syphilis cases in the country were reported from only twenty-eight counties, mostly in large cities and in the South, making it easier for the CDC to focus and target their efforts.

One of the CDC's prime targets was the city and county of San Francisco. In the 1970s and '80s, San Francisco had been saddled with the notoriety of leading the nation in new cases of syphilis, reporting more than two thousand cases a year. By 1998, the year that Klausner became the city's STD controller, there were only forty syphilis cases the entire year. Only eight of those cases were among men who had sex with men, the lowest number ever recorded in the city.

In Mark Dworkin's book *Outbreak Investigations Around the World*, Klausner described his hopeful mood during this time. "It was perfect timing to eliminate syphilis. All the conditions required for potential disease eradication were in place—only human-to-human transmission, low case counts, an easy and accurate test, and highly effective treatment [penicillin] that both prevented and cured infection."[12]

He formed a syphilis elimination team with his internal staff and members of various community-based organizations. They drafted a response plan and met biweekly to review and improve their current efforts. Klausner reviewed every new case personally to make sure they'd been adequately treated. He exercised constant vigilance; every case could represent an outbreak that could hamper their elimination efforts.

In June and July of 1999, he received reports of two cases of early-stage syphilis (acquired in the previous year) in gay men. Recalling his interaction with Bill a few months before, he had told his contact tracers to ask specifically about meeting partners on the internet, which websites they used, then where they subsequently met to have sex—a house, a public place, bar, nightclub.

It turned out both men with syphilis had met most of their sex partners in the same AOL chat room frequented by Bill, SFM4M. Here was the rub: they didn't know the names of most of their partners, as they only identified themselves by an online handle, or screen name. Klausner asked his DISs to go back and reinterview everyone diagnosed with early-stage syphilis in 1999. All but one person who had used the internet to meet sex partners had used the SFM4M chat room. Klausner asked each DIS to gather as many online handles as they could in the hopes that they could later be used to discern the structure of the sexual network.

Based on the results of their investigation, Klausner and his team were able to construct a network of interconnected cases of syphilis. All cases were linked through the chat room, and a few reported dozens of partners in the six months prior to diagnosis.[13] Once they'd identified the cases and confirmed that the outbreak was linked to a virtual place in cyberspace, they found themselves in uncharted territory. How could they warn the other users of the chat room what was going on? None of Klausner's colleagues or his contacts at the CDC knew what to do.

Klausner tried to contact AOL. His hope was to provide the screen names that his investigators had collected, and in turn AOL would provide him with the contact information of their users who had been exposed: name, address, email, phone number, and so on. With that information, he could have his investigators contact the other chat room users, notify them of their potential exposure to syphilis, and encourage them to be tested and treated. Perhaps AOL would collaborate with him to create a more sexually healthy chat room environment, promoting regular STD and HIV testing and posting links to information on safe sex practices.

There was one problem with Klausner's plan. No one at AOL headquarters would return his calls.

As news of the syphilis cyberspace outbreak began to reach the local media, Klausner finally reached a public relations executive at AOL, who informed him that they would not share the private information of their users with him, even in the interest of public health. Klausner would not be deterred. Through a friend's sister, he contacted a corporate attorney for AOL in Virginia, sending her emails, faxes, and certified letters making sure she was aware of the serious nature of the situation. Through another colleague, he reached out to a gay senior executive at AOL Time Warner in New York, thinking he might be sympathetic to issues related to gay men's health.

Finally, AOL corporate headquarters informed him that "such personal contact information of AOL members could only be released on federal subpoena or in the case of an imminent grave danger—such as homicide or suicide." He doubted that he'd be able to get a subpoena, and he couldn't make a case that AOL members were in grave danger. In the post-penicillin era, syphilis was easily treatable and almost never fatal.

He reached out to the CDC, sending messages to the Division of STD Prevention and higher up the chain to the director of the National Center for STD, HIV, and TB Prevention. While his colleagues were personally sympathetic, they did not feel that the CDC should officially intervene with AOL Time Warner and did not want to make a federal case out of it. Klausner was dumbfounded. This was currently only happening in San Francisco, but tomorrow, it would happen somewhere else. Couldn't they see what was coming?

Finally, he got an unexpected break. Perhaps AOL realized that he was a headache that was not going away or sensed a potential public relations disaster. They introduced Klausner to the CEO of PlanetOut, an online community for the LGBTQ community that was supported by AOL. PlanetOut's CEO immediately sprang into action. He rallied a dozen of his employees, including programmers, writers, and marketing staff, to log on to the SFM4M chat room every day for two weeks. They informed users about the outbreak, offered syphilis education, and encouraged chat room users to get tested. Based on chat room usage patterns, Klausner estimates that thousands of AOL users were informed of the outbreak. Through PlanetOut, Klausner surveyed thirty-five users after their two-week intervention; two-thirds agreed that PlanetOut's syphilis awareness campaign in the chat room was useful and appropriate.

While he was pleased that PlanetOut had reached so many users, Klausner still had an ax to grind with AOL. Through the San Francisco Department of Public Health, he issued a press release in August of 1999, linking the chat room to the increase in syphilis cases. He publicly shamed AOL for refusing to inform chat room users themselves. His goal was twofold: he hoped to receive local media coverage that would raise syphilis awareness in San Francisco, and national media attention that might cajole AOL into acting.

While he never got the response he wanted from AOL, he had inadvertently begun a national conversation about individual privacy versus the duty of government to protect the public's health. Following the press release, the syphilis outbreak ended up as a front-page story by Evelyn Nieves in *The New York Times*: PRIVACY QUESTIONS RAISED IN CASES OF SYPHILIS LINKED TO CHAT ROOM.[14]

The AOL chat room outbreak was a wake-up call for public health departments, who were woefully unprepared to conduct contact tracing using the internet. Even in 1999, some state and local health departments did not regularly use email or have access to sites outside their own internal web pages. Klausner's staff in San Francisco went on to create protocols and procedures to notify partners of an STI exposure over the internet. A new discipline within the field of STI control was born: internet partner services, which evolved from notifying partners on message boards or email, to texting and messaging on smartphone apps.

As for syphilis, the rise of the internet had helped release a genie from a bottle. The number of syphilis cases among gay men in San Francisco would rise from 8 cases in 1998, to over 550 in 2004, to over 1,500 in 2019. Nationally, there were more than 37,000 cases of primary and secondary syphilis (the most infectious stages) in 2019, and almost half of those were in men who have sex with men.[15] It seems the genie will not be back in the bottle anytime soon.

Despite his beef with AOL, Klausner never bemoaned the rise of the internet to facilitate sex. Did this profound shift in sexual ecology make his job more difficult? Sure. But at his core, he was a sex-positive pragmatist. He took the chat room–based hookups in stride as just one more thing to navigate in the world of STD control.

"We want people to be sexually active, we want people to hook up, we also want people to practice safer behavior and have access to testing and screening. So I've been trying to recognize that this is a new, efficient way for people to fulfill an important part in their lives," he said. "Every new successful technological advance—printing press, photography, telephone, cinema, television, and the internet—seems to have one thing in common: its ability to enhance life's sexual experiences."

It's easy to see why people go online to get laid. Hooking up without technology can be quite labor intensive: showering, maybe doing hair and makeup, changing out of sweatpants, going to a bar/party, making conversation, and bringing someone home. The AOL chat rooms helped eliminate some of that effort so hooking up became easier—though at the same time syphilis and other STIs began to flourish once again, and tracking sexual contacts became more difficult.

Despite Klausner's foresight about the internet's power to facilitate hookups, he couldn't possibly have predicted smartphones and the explosion of hookup apps. According to the Pew Research Center, 95 percent of U.S. adults own a cell phone today, and 77 percent of those are smartphones.[16] Looking at two among the dozens of dating apps, Grindr (targeting gay, bisexual, trans, and queer people) has over three million daily users, while Tinder has been downloaded one hundred million times, with ten million active users per day.[17,18]

With apps making online hookups easy, free, and available to all, how could the sex detectives possibly keep up?

A Face Without a Name

Charles Fann could tell that this case was not going to be easy. It wasn't the index patient's fault—he had just been diagnosed with syphilis and was willing to help Fann track down his sex partners over the past three months. The patient knew most of his partners well—they would be easy to find. But for one of them, the only identifying information the patient had was an online handle (SFJoJo) and a profile picture from a dating app.

This was a familiar story to Fann, a DIS supervisor for the city of San Francisco from 2012 to 2016. He mostly oversaw the work of the junior DIS, but he'd occasionally work up a case himself to keep his skills sharp. A trim, Black gay man in his early thirties, his clients didn't always realize that he was an official of the city health department. To counteract his youthful appearance, he was meticulous with his attire. His button-down shirts were always ironed and tucked; his dress pants were always pressed.

Fann stumbled into this career path by happenstance. In his early twenties, when he was still in the closet and living in Tacoma, he decided to venture into a gay bar. There, he was approached by an outreach worker passing out condoms. Initially, he was skeptical of the worker's intentions: "Do I *look* like someone who needs a condom?" But he accepted a package of condoms, flipped it over, and was intrigued to see an invitation to a "condom party," which turned out to be a focus group for gay men, hosted by a local nonprofit organization. After the focus group ended, he started volunteering for the nonprofit, doing outreach to increase condom use among adolescents and young adults in Tacoma. He bumped into a local DIS during the course of his outreach, assisted him to work up a case, and was sold.

Shortly thereafter, Fann left Tacoma and headed for the gay mecca of San Francisco. To a bystander, he resembled one of the thousands of San Francisco singles out on the prowl. But unlike those seeking sexual awakening or adventure, Fann had a very different conquest in mind. He came to join the ranks of a dozen DISs trying to locate people exposed to tuberculosis, HIV, and STIs all over the city.

Once he arrived, he quickly learned every neighborhood, park, bar, nightclub, and gym in the city—any location where a contact might be found. Once he located who he was looking for, he'd persuade them to do a contact tracing interview. He'd conduct these interviews in the STI clinic, in people's homes, or even in his car.

If he couldn't physically find someone, he'd search for contacts online. To do this, Fann would use the standard sites you or I might use: Google, Facebook, Instagram. As a DIS, he could also access DMV records, databases of public records, and information from commercial data brokers. But all these searches relied on one thing: knowing the name of the person he was looking for.

Finding someone was trickier when you had just a picture and an online handle from which to work. Online handles could potentially be traced, but it was clear by this point that website and app owners weren't about to give up that information. When these cases arose, Fann encouraged his staff not to give up right away. "Instead of focusing on what you don't have, how can you analyze what you do have?"

Fann looked at SFJoJo's profile picture, which showed a fit and trim guy in workout clothes standing in front of a building. The photo was angled to show the signage for a local gym. The gym had several branches in the city, but Fann recognized the location based on the appearance of the street and the surrounding buildings. He made a simple deduction: a man in workout clothes in front of a gym probably frequents that gym. It might not be correct, but it was definitely a start.

Then Fann noticed something else in the background of the picture. In the distance, there was another man walking inside the gym's front door, with a button-down shirt on a hanger slung over his shoulder. The man in the background was probably going to change into this shirt to go to work after exercising, Fann thought. If that was the case, SFJoJo had likely taken this picture in the morning. Perhaps mornings were his preferred workout time. Now they knew where and roughly what time of day they might be able to run into him.

Fann sent his team member Erin to the gym to stake it out. Prior to Erin's arrival, Fann had called the gym to let them know that the health department was looking for a gym member who'd been exposed to a communicable disease. They asked to use one of the gym's offices so Erin could have a private meeting if needed. Fann had also asked if they could set up a table with informational brochures so gym members could mingle with Erin as they wandered in. They decided that the materials would be about a topic unrelated to sex: in this case, helmet safety.

Erin set up the table around 7:00 a.m., including placing stacks of materials about the city's helmet safety program. Gym members approached the table throughout the morning to take brochures and chat. She also informed the members about other health services offered by the health department, including the city's clinics for tuberculosis, travel vaccines, and STIs. All the while, Erin scanned the steady stream of gym-goers looking for SFJoJo.

After three hours at the table, she spotted SFJoJo entering the lobby. She asked if she could speak to him and pulled him into an office. After she had verified his identity with his picture and his handle, she told him he'd been exposed to syphilis. He was stunned; this was the last thing he was expecting on his way to the gym that morning. He was grateful that the health department had found him. Skipping his workout, he walked five blocks away to the STI clinic to be tested and treated.

Fann was modest about his team's success in this case. "People thought I had this psychic thing going on, but it really wasn't that. It was just basic investigative work." He presented the case of SFJoJo to other DISs around the state. Some of them complained that the handles and pseudonyms on dating apps made finding contacts impossible. "I don't know how to counsel or interview someone [a patient] who has sex on Grindr." Fann would correct them. "No one has sex on Grindr, because it's not a place. You still have to have a place." His point being, if they could figure out something about the place, it might result in useful information about the person or people involved. Missing someone's traditional contact info—a name and phone number—might not equate to a dead end.

Fann felt that tenacity was the key to success. If an index patient said, "You'll never find him because I had sex with the guy in his car," Fann encouraged the DIS to try to coax out more information. "Well, what do you remember about the car? What color was it? Was there anything hanging from the rearview mirror?" Then the person might remember how "he had a residential parking sticker from X neighborhood on his bumper." Now there would be something to go on.

Performing contact tracing online has gotten easier for some health departments. Some counties allow DISs to create profiles on the most popular dating apps. Once a DIS has someone's online handle, the worker goes onto the app and messages the partner directly from the health department's account, letting them know that they were exposed to an STI. The profile picture they use is often an official logo, but some apps require a photo of an actual person. Imagine if you were to open your dating app to look at your messages, see a profile picture and expect an invitation to hook up. Instead, it's the health department wanting to treat you for syphilis.

As one client summed it up for Fann: "I thought it was a bad joke when I received a message from the health department on this site. Why would the health department be on this site and how did they know my handle? I was angry when I called the counselor who messaged me. He was very patient with me and explained to me why he was contacting me. He said that he could talk to me in person if I preferred and even offered me free confidential testing at the clinic he worked at."[19]

Getting news of an STI exposure from your dating app could be jarring, but it would probably motivate you to do something about it. David Katz from the University of Washington surveyed men who have sex with men who used apps to hook up. While 70 percent preferred hearing the news directly from their partner(s), almost everyone was willing to get tested if messaged about an exposure by the health department (95 percent) or an anonymous in-app message (85 percent).[20]

No matter who delivers the news, they have to be sure they are contacting the right person. The handle hotguy4U and hotguyforyou likely belong to different people on the same app. Some apps allow multiple profiles with the same profile name if they are in different states, making it difficult for a DIS to be sure that they are reaching the right person.

There's no question about it—the geosocial dating apps (e.g., Grindr, Plenty of Fish, Tinder) are great if you want to get laid and are feeling slightly lazy. They utilize GPS to improve the speed and efficiency of finding partners by letting you select not only what you are looking for sexually but how far (geographically) you are willing to go to get it. The eligible partners you see will depend on where you were when you opened the app and began searching.

But that convenience comes with a cost. Weeks later, if you are diagnosed with an STI, your partners may be nearly impossible to track down. If someone turns off their app or they've moved out range from where you are, their profile may no longer be visible to you. If you don't have a name or number, a record of in-app messages, or their online handle, you may not have a reliable way to find them later.

For John Potterat, tracing partners in 1970s Colorado Springs, there were limited physical venues where people were likely to meet. "We keep running into the same people, they keep going to the same addresses, the same places of social significance—bars, clubs, whatever." But in today's landscape, with geosocial dating apps, you can arrange a hookup wherever you happen to be when desire strikes: Starbucks, public transit, a university lecture hall, a church parking lot. Now there are infinite places to cruise for partners, and the brick-and-mortar places of "social significance" may not be as significant anymore.

The difficulty of contact tracing in the age of apps is reflected in the inability of DISs to get patients to name partners and to get those partners treated. Rilene Chew Ng, an epidemiologist for the City of San Francisco, has studied partner services for more than ten years. She's heard from DISs who say, "Oh, we're just not pushing hard enough to get names, we're not trying different strategies, we give up too soon." On the other side are those who say, "Well, they can't name their partners, so I'm not going to keep pushing."

Her team decided to interview men with syphilis, both those that agreed to have partners notified by public health workers and those that refused. They were trying to get a sense of how people were meeting each other, what apps they were using, and how much information people actually got on their partners.

As Chew Ng pored over the transcripts, she began to feel discouraged at what the men were saying. There was everything from, "I deleted this person's profile . . . they were blocked from my app," to "There's just no way to find this person. If I see them pop up in my app again, sure I can give that information to you, but I genuinely don't know how to get ahold of this person."

By 2017, over half of patients in San Francisco with new syphilis diagnoses met their partners on geosocial dating apps, while those using computer-based websites declined to less than 30 percent. At the same time, less than half of patients with syphilis in San Francisco named at least one partner for the health department, and only one in four had at least one partner who was treated.[21] With the cases piling up and a minority of patients naming partners, are contact tracers fighting a losing battle?

According to Chew Ng, "If the goal is to find the newly infected folks and bring them to treatment, then no, I don't think that partner notification is really doing what we want it to do. It's very difficult for the DIS because there is this whole spectrum of work but the only outcome we talk about is, did they get partners' names or not? And often that's out of their control." She admitted, "Unless we intervene drastically to change how we try to find and contact people, it's just not going to be the intervention that it used to be."

Matthew Golden, director of STD/HIV prevention for Seattle/King County in Washington, thinks we should change the goal of partner notification. Success shouldn't just be defined by how many sex partners can be named and found. Rather, it should be about trying to figure out "what can be done on this person's behalf that will advance their own health and advance the health of our community?" He argues that we should be thinking about things more holistically, especially when there's an opportunity to interact with a high-risk population of people. Because syphilis and HIV often go hand in hand, Golden feels that an opportunity to intervene with syphilis is also a chance to address HIV prevention and care.

"I think partner services for syphilis is an excellent way to promote use of HIV pre-exposure prophylaxis [PrEP] . . . and in some places is probably a good way to identify people who are HIV-positive and out of medical care." But given the challenges of naming and finding partners in the internet age, he admits that partner services likely has only a modest impact on syphilis.

In today's age of apps, there is national debate over whether overwhelmed health departments should continue to notify partners about STIs, particularly men who have sex with men, who account for a large proportion of cases. Supporters argue that the individual-level benefit of treating patients and partners is worth it. Detractors say partner notification is not effective at reducing infections at the community level (although there aren't a lot of alternatives). Even though partner notification has been around since the 1930s, the practice has not been vigorously studied. A 2013 review by Cochrane concluded that there wasn't enough evidence to say whether partner notification by health departments was actually effective at reducing STIs.[22]

Then there is the issue of funding. According to David Harvey, director of the National Coalition of STD Directors, "Federal STD funding [in 2018] has seen a 40% decrease in purchasing power since 2003. That means that state and local health departments, most of which depend primarily on federal funding to support their STD programs, are working with budgets that are effectively what they were 15 years ago."

Compared to 2003, there are fewer DISs on the ground today, and finding partners is more difficult than ever before. According to Golden, "We clearly have a mismatch between the resources we have available for partner services and the scale of our problem."

As the STI epidemic continues to escalate, we're in urgent need of scrappy sex detectives on the front lines. Although the work is challenging and flawed, it is the best solution we have until something better comes along. But should DISs keep putting boots on the ground to search for sexual contacts? Should they concentrate on working virtually, leveraging dating apps to reach index patients and their partners? The future of partner notification for STIs remains to be seen. Regardless of what method they choose, today's contact tracers must hoist up their boulders and gird their loins; the mountain is becoming ever steeper and harder to climb.

7

A Pox on Both Your Houses

On the Mysteries and Maladies of Syphilis

Making Syphilis Great Again

I hear bloodcurdling screams emitting from the bowels of my home.

It's a familiar tale, two brothers locked in battle over the same possession. There is no logical reason for this, as there are so many toys in my basement-cum-playroom that each could play with something different every day for weeks without having to share. But logic has no place when two kids share the mindset of "that's mine, give it back."

The screams continue. Fearing Cain-and-Abel redux, I stick my head down the stairs. Two boys (ages eight and three) are fighting over a stuffed animal, a pink, curly worm with beady black eyes. A closer look at the tag revealed its identity: "Syphilis," printed in six languages. This tussle of wills is over a stuffed STI, enlarged to one million times its usual size.

"*Mom!!!* Zane has syphilis, and he won't give it to me!"

"Dat's mine! You hab herpes!"

"I don't want herpes. I want syphilis!!"

"Well, you're mean! You need to share wid me!"

"*No!!* It's *mine*!!"

(Wailing ensues.)

Technically, these toys are mine. I purchased several STI stuffed animals and mounted them to my office wall like faux hunting trophies. They were promptly stolen when I took my kids into work one day and became the darlings of the playroom and beloved bedtime companions. Now I live in fear that my kids will tell people that their mother gave them syphilis, gonorrhea, and herpes.

Begrudgingly, my older son relinquished the stuffed syphilis to his brother. But so precious was syphilis that I had to promise him a replacement.

"How about another syphilis?"

"Yes. I also want norovirus, or maybe *E. coli*. Could Santa bring them for Christmas this year?"

"Yes, I'm sure Santa could arrange for that."

My kids are the only two who seem to want syphilis. Everyone else I've diagnosed has expressed negative feelings toward the infection: disappointment, dismay, even shame.

Syphilis is thought to be one of the oldest *venereal diseases,* an umbrella term that also includes gonorrhea, plus rare ulcer-causing STIs such as donovanosis, lymphogranuloma venereum, and chancroid. The moniker alone, *venereal,* bears negative associations with venery, or sexual indulgence. And syphilis, the most disfiguring and deadly of those venereal diseases, was the most stigmatized of the lot.

Shame and stigma surrounding syphilis goes back for centuries, starting with the fact that no one will take credit for its introduction into the population. When the first recorded outbreak of syphilis swept through Italy in 1495, the Italians were quick to blame it on the French invasion, widely referring to it as the *Mal Francese*. When it hit Spain, Christopher Columbus's physicians blamed the Native Americans in the New World. By 1500, syphilis had spread throughout western Europe to the Scandinavian countries, Poland, and Russia. European explorers brought it with them to India; by 1520, it had reached Africa, China, and Japan.[1]

As syphilis spread throughout the world, each country's populace changed its name to blame the epidemic on someone else, often one of their enemies. The English joined the Italians in calling it the *French disease*, the Russians called it the *Polish disease*, the Indians blamed the Portuguese, the Tahitians pointed their fingers at the British, the Japanese called it *Chinese pox*. Syphilis was a hot potato that no one wanted in their laps.

Regardless of who was to blame, the initial epidemic of syphilis in Italy certainly wreaked havoc on the population. The exact death toll is unknown, but in *The Great Pox*, historian Jon Arrizabalaga describes how the overwhelming number of cases in Italy necessitated the creation of an entire hospital system for the *incurabili*, those with incurable syphilis who were poor and rejected from traditional medical centers.[2]

The sheer number of cases was only part of the issue. Symptoms described by historians during the *Mal Francese* outbreak were much severer than those observed today. Italian clergyman Ser Tommaso di Silvestro left behind a detailed written account of his personal experience with the disease. In *The Great Pox*, he describes the extent of his suffering: "And all my head became covered with blemishes or scabs, and the pains appeared in my right and left arms . . . from the shoulder to the joint with the hand, the bones ached so that I could never find rest. And then my right knee ached, and I became covered in boils, all over my front and back."

More than five hundred years after the first outbreak of the Great Pox, the intensity of its symptoms has declined, but its complexities remain unchanged. Syphilis is known in medicine as the *great imitator;* it is notorious for mimicking hundreds of other conditions. It can look like almost anything, or nothing at all. The father of modern medicine, Sir William Osler, appreciated how difficult it was to understand the disease. His guidance is often quoted to clinicians in training today: "Know syphilis in all its manifestations and relations, and all other things clinical will be added unto you."[3]

During Osler's time, it was easier for physicians to know syphilis because there was simply more of it around. In a speech to the Medical Society of London, Osler estimated that in 1915, syphilis accounted for approximately sixty thousand adult deaths and fifteen thousand to twenty thousand infant deaths in the UK.[4] There were tools to control the spread of infection, including a blood antibody test (the Wassermann test) and effective—if imperfect—treatment. The drug salvarsan was introduced during this time, which was a huge improvement over the toxic mercury used for treatment in years past. Although it did provide a cure, it also was based on arsenic, causing some users to suffer liver damage and death. Plus, salvarsan was costly and required repeated treatments for up to a year.

Treatment wasn't even the biggest hurdle. According to Osler, the greatest barrier to addressing syphilis was that "centuries of silence had made venereal disease taboo." Osler felt the stigma was so deep "that patients avoided hospital and even their family doctors." This secrecy meant that getting an accurate case count and death toll was impossible. He admitted having to extrapolate his death counts from syphilis because he knew that they were grossly underestimated. Like the early days of the AIDS epidemic, the cause of death for a patient with syphilis was often attributed to anything but the actual infection that caused it. As Osler put it, "Syphilis had been and remained the despair of the statistician. Even in death a stigma was associated with it, and the returns were everywhere but under the special caption of the disease itself."

Osler realized that there would be no winning against syphilis until there was an honest accounting of its impact on the population: "To be successful in any fight the primary essential was to know where the enemy was placed." Unfortunately, he would not live to see his wish fulfilled. Despite shifts in sexual mores in the 1920s and greater openness to sexual expression, this did not translate into more open discussion of STIs, prevention, testing, and treatment. The scourge of syphilis was able to continue under the public's radar largely unnoticed.

It wasn't until 1936 that the veil of ignorance was lifted from the public's eyes. By that time, the U.S. Public Health Service estimated that one in ten adults harbored a syphilis infection, and Surgeon General Thomas Parran decided it was time for the world to take notice of syphilis. In July of 1936, he published an article entitled "The Next Great Plague to Go" in the periodical *Survey Graphic,* which used pictographs to describe rates of syphilis infection in the population and the devastating outcomes of untreated syphilis in pregnancy. *Reader's Digest* followed, publishing another version of his article, "Why Don't We Stamp Out Syphilis," spreading Parran's message to its five hundred thousand subscribers.

Parran's campaign gained momentum from there. According to medical historian Allan Brandt, Parran placed additional articles in 125 major newspapers all over the country.[5] In October 1936, his anti-syphilis crusade landed him on the cover of *Time* magazine. Under Parran's picture on the cover, there was no mention of syphilis or venereal disease. Instead, there was a vague reference to his goal: "His target is behind a taboo."

Parran then joined forces with science writer Paul de Kruif, setting his sights on the genteel readers of the *Ladies' Home Journal*. Although his article was rejected at first, the *Ladies' Home Journal* relented in August of 1937, publishing "We Can End This Sorrow" and reaching their readership of three million women.[6] He capped it off with his own bestselling book, *Shadow on the Land*, summarizing the themes of his anti-syphilis campaign: testing, treatment, and raising public awareness. "Find syphilis, treat syphilis, teach syphilis" was his mantra.[7]

In 1938, after passing in both the House and Senate, President Franklin Delano Roosevelt signed the National Venereal Disease Control Act into law. Federal funds became available for research, construction of public STI clinics, and provision of free drugs to private-sector physicians. Premarital and prenatal syphilis testing laws went into effect in many states. The country had finally woken up to the problem of STIs and was mobilized to act.

At the time, syphilis was prominent enough to warrant its own specialty within the field of medicine. Given that syphilis often manifested with skin ailments, the dermatologists decided to take ownership of it. In 1936, they named their professional licensing board the American Board of Dermatology and Syphilology.[8]

Some dermatologists were not thrilled to be associated with an STI. Three years after the founding of the board, notes from their annual meeting revealed some of their members' discontent: "An objection has been raised by one of our diplomates . . . although he values his certificate and would like to display it in his office, he is unwilling to do so unless the word 'syphilology' is either deleted or possibly printed in small type."

The chagrin of the dermatologists toward syphilis would continue; similar debates occurred at their board meetings for almost twenty years. In 1953, the American Board of Dermatology and Syphilology decided to drop *syphilis* from its title, as did the American Academy of Dermatology and Syphilology. None of the other medical specialty boards rushed in to claim syphilis as their own. Syphilology became a lone man with no country.

In the United States, Europe, and India, the field of syphilology eventually expanded and became known as *venereology*, a term more inclusive of other sexually transmitted diseases. Some questioned whether classifying syphilis as a venereal disease was fueling qualms about discussing it openly. Should syphilis, which could be transmitted in utero or between two lawfully married people, be associated with such an unsavory term?[9]

It wasn't until the 1970s that *venereology* and *venereal disease* would be replaced by *sexually transmitted diseases* (STD). Regardless of what it was called, the field of venereology would remain a low member on the totem pole of medical specialties. In the satirical *Bluff Your Way in Doctoring*, author Patrick Keating provides an accurate tongue-in-cheek description of the field's lowly status within medicine:

> Venereologists are in the main general physicians, who, for reasons best known to themselves, specialise in the troublesome ailments of the adult reproductive and plumbing systems. The layman will refer to this taboo area of expertise as the "Clap Clinic" in the saloon and the "Special Clinic" in the lounge. Either way, it is spoken of in hushed tones and nobody likes to have to go there . . . If you are a pillar of the church, have led a sheltered life thus far and think that "intercourse" means the same thing as "conversation" then this may not be the area for you.[10]

The unmentionable subject matter and dubious standing within organized medicine would keep venereology out of the academic mainstream in the United States until 1968. That was the year that a soft-spoken infectious disease physician named King Holmes arrived at the University of Washington, determined to make STIs respectable once again.

They say that men can father offspring well into their seventies, and Holmes is no exception. He is widely regarded as the patriarch of contemporary STI research in the United States. Beginning with seven postdoctoral researchers five decades ago at the University of Washington, his descendants now number in the thousands, scattered throughout academia, public health, and clinical medicine.

For nearly thirty years, Holmes has hosted a two-week-long boot camp on STIs and HIV at the University of Washington, which has graduated more than twenty-five hundred scientists worldwide (myself included). At the end of the boot camp, everyone receives a copy of his massive two-thousand-page textbook, *Sexually Transmitted Diseases*. One lucky attendee also wins a copy of one of his other books, cowritten with graduate student Jennifer Wear: *How to Have Intercourse Without Getting Screwed*.

How to Have Intercourse was published in 1976, during a boom time for STIs in the United States. Back then, Holmes admitted that "despite all the efforts of sociologists and venereologists, no one yet fully understands why STD is booming. The number one reason may be sexual liberation which followed development of newer methods of contraception" (e.g., the Pill and the IUD).[11] When it came to syphilis, however, Holmes recognized that the epidemic was concentrating among men who had sex with men, who were enjoying a sexual liberation of their own in cities such as San Francisco, Los Angeles, and New York.

When HIV/AIDS struck in the 1980s, it would take several

years for people to get wise to the magnitude of the disease and alter their sexual behavior. Eventually, fear of death from AIDS caused people to reduce their number of sex partners and increase their use of condoms. As a result, during the late 1980s and early 1990s, rates of syphilis and other STIs began to decline.

But it wouldn't last forever. By 2001, cases began climbing again, and by 2018, there were more than 115,000 cases of syphilis in the United States, and over half of these were among men who have sex with men.[12–14]

Why did syphilis surge once again? Holmes remembered being on the AIDS Walk in 1996, when he strolled by a doorway displaying a local newspaper aimed at Seattle's LGBTQ population. On the front page was a story heralding the medical breakthrough of highly active antiretroviral therapy for HIV. Above the story there was a photo of a smiling young man with a caption: "Now I don't need to worry about condoms anymore."

Holmes reflected, "We had declining rates of everything, because people were really scared of dying. Then all of a sudden, we had a highly active antiretroviral therapy and people weren't scared. That's part of what happened. And now we have HIV preexposure prophylaxis [a.k.a. PrEP—more on this in chapter 9], which is quickly changing sexual behaviors. That is happening on a global basis." He recalled a recent argument he had with a colleague at the World Health Organization about the most likely culprit for the current resurgence in STIs. He soon realized it didn't matter which one of them was right. "The real issue is how are we going to deal with this. Not just in the States, but everywhere."

As syphilis continues to increase at an alarming pace, it doesn't matter which medical specialty claims ownership for it, or which of many factors are to blame for its return. There is no time for shame and stigma amid an epidemic. Following Thomas

Parran's advice from the 1930s would still help disrupt disease transmission: find syphilis and treat syphilis, the earlier the better. Unfortunately, finding it early is not always easy; sometimes you must fumble in darkness before you see light.

Keeping Your Eyes on the Ball

Everyone was excited when the lights went out.

At the end of the narrow laboratory in San Francisco City Clinic, physician Joe Engelman had shut off the overhead lights as he sat down at the laboratory bench. A single dim light from a microscope illuminated his grizzled beard from below. The faded print of his Grateful Dead T-shirt was just visible in the shadows, musing WHAT A LONG, STRANGE TRIP IT'S BEEN. Hunched just behind him were the silhouettes of a nurse, doctor, and medical student. Each was poised in anticipation, hoping for a glimpse of the elusive *Treponema pallidum*, the spirochete bacterium that causes syphilis.

The specimen under Engelman's microscope was from a gay man in his twenties, who had come to the clinic because he had developed a painless ulcer on the shaft of his penis. During the physical examination, Engelman had cleaned off the ulcer with some moist gauze, then firmly squeezed the ulcer to produce a drop of fluid. Placing the fluid on a glass slide, he promptly headed to the laboratory to examine it under the microscope.

Because Engelman had suspected a diagnosis of syphilis, he did not use the ordinary microscope, which shines a bright beam of light up through a specimen on a glass slide. While this works for visualizing cells and some bacteria, *Treponema pallidum* is exquisitely fine and difficult to see. This is where a dark-field microscope comes in. It condenses the light into a beam, then scatters it obliquely through a drop of oil placed against the bottom of the slide. Performed in the dark, the scant light produces an image like an x-ray: bright white against a black background.[15]

Dark-field microscopy is a dying art. As far back as 2009, a study of U.S. infectious disease specialists found that only one in ten had ever used dark-field microscopy to diagnose syphilis.[16] Most didn't even have access to the test. Today, it's limited to a smattering of specialty clinics and academic centers throughout the United States.

All the stars must align to perform a dark field successfully. The patient must show up with an ulcer moist enough to produce fluid for the test. The fluid must be examined immediately to visualize the bacteria while they are still alive. But when conditions are right, there are few things as satisfying as seeing the tiny white corkscrews of *Treponema pallidum* twist and undulate against a starry black background, like fusilli pasta swimming in squid ink.

Dark-field microscopy is one of two ways to definitively diagnose syphilis. Methods like blood antibody testing are easier, but don't always give a direct answer (more on that later). The only other definitive test requires extreme measures. Until recently, *Treponema pallidum* could not be cultured in a laboratory, so the only way to grow it was to take infected tissue from a person who is suspected of having syphilis, inject it into an animal, and see if the animal develops syphilis. *Treponema pallidum* will grow in a variety of animals from hamsters to chimpanzees, but it's the rabbit that serves as the best animal model to test for syphilis in humans.

When I first heard about this, I wondered two things: How did syphilis get from the human to the rabbit in the first place? Then once infected, what did the rabbits do? I had visions of lustful syphilitic bunnies having their way with each other while disinterested scientists looked on. But according to syphilis researcher Sheila Lukehart at the University of Washington, rabbit infectivity testing for syphilis is not salacious in the least, and it involves little pleasure for the rabbit.

First, a sample from a human (fluid from an ulcer or spinal tap) is injected directly into one of the rabbit's testicles. Following the injection, some unlucky laboratory worker is assigned the role of testicle watcher. Within a week, the injected testicle may begin to swell and become inflamed, a sign that syphilis was present in the human sample.[17]

Sometimes all remains quiet, so the testicle watching must continue. The rabbit's blood is tested every month to see if it has developed antibodies to syphilis during this time. If the rabbit's testicles remain quiet and it doesn't develop antibodies to syphilis after three months, the test is considered negative.

If the rabbit develops antibodies, but the testicle doesn't swell, the rabbit is euthanized and castrated. Then the first rabbit's testicle and lymph nodes are ground up, injected into a second rabbit's testicle, and the watching begins again. If this second testicle swells, this confirms the presence of syphilis.

Rabbit infectivity testing for syphilis is very accurate, but clearly it's expensive and impractical. Imagine asking your doctor for a syphilis test, only to be told, "Give me two rabbits and three months, and I'll get back to you." Thus, rabbit infectivity testing is only used for research purposes and is only performed in a few U.S. universities, including Lukehart's laboratory.

Today, syphilis is typically diagnosed using two types of blood antibody tests, but they are rife with issues. One type of test detects specific antibodies to *Treponema pallidum*, but the antibodies stay around for life, even after the infection is cured. It's possible to be infected with syphilis multiple times over one's lifetime, but with the treponemal antibody test it's impossible to distinguish a new infection from an old one.

A second type of test detects antibodies to cholesterol or lecithin that are created by cellular damage from syphilis; the most common of these is the rapid plasma reagin (RPR) test. But this test can be reactive in a host of other conditions, including hepatitis, cancer, HIV, or autoimmune diseases such as lupus. The RPR test also misses one in four cases of primary syphilis, which is syphilis at its most infectious stage. Without a smoking gun like a positive dark field or rabbit infectivity test, a clinician must piece together the diagnosis of syphilis using clues from the patient's sexual history, their physical exam, and a combination of imperfect antibody tests.

That day, Joe Engelman was in luck. His specimen at the microscope was teeming with *Treponema pallidum*. He nodded and moved over to make room for the others to see. Contented sighs could be heard as everyone took a turn at the microscope. This positive dark field test, combined with the ulcer on the patient's penis, meant the diagnosis of syphilis was a certainty.

Textbook cases of syphilis like the one on Engelman's microscope are often not the norm. Syphilis can imitate so many different conditions yet is still rare enough that clinicians can go years without seeing a patient with symptoms. Making the diagnosis can be worse than looking for a needle in a haystack, because in this case, the needle can potentially take any size, shape, or form— or even be invisible.

So how are most of these diagnoses made? A high index of suspicion, even a touch of paranoia doesn't hurt. With syphilis presenting in many shapes and forms, almost any neurologic or dermatologic ailment in a sexually active person could potentially be a sign. Ears ringing? Rash on the chest? Feeling tired? Could be syphilis. Add in a good sexual history and antibody testing, and it can help distinguish whether something is syphilis versus hundreds of other conditions that might look the same.

Then sometimes there are telltale clues on the body, times when *Treponema pallidum* leaves a trail of bread crumbs right to the diagnosis. Those instances require testicle watchers to be at the ready once again.

Setting the Stage

What are you doing down there?" asked Daniel, my twenty-eight-year-old patient with black briefs around his knees. He was peering down at me as I was crouched on a stool, lifting his penis with one hand and craning my neck down to look under his scrotum. He looked amused by my gymnastics.

"I'm looking to see if you might have a rash from syphilis."

"Really? On my balls?"

"Yes. Syphilis can show up on your hands and feet, scrotum, or your eyes. It can appear almost anywhere in your body."

Daniel had never had syphilis, and it wasn't on his radar when he walked into the clinic that morning. He'd come in because he had started having unprotected sex with a new partner about three months earlier. Things were going well in their relationship, but last week Daniel had begun feeling tired and run-down; he was anxious that he might have contracted HIV. Syphilis can cause many of the same symptoms as a recent HIV infection: fatigue, muscle aches, rash, swollen glands.

I finally maneuvered into position and saw two raised pink oval spots on his scrotum, each just three centimeters across. Perhaps a contortionist could have self-identified these spots, but Daniel would likely have needed a mirror to be able to see them easily. In addition, the lymph nodes in his groin were swollen, but there was nothing unusual on his penis or the rest of his skin.

A mantra from Joe Engelman began to repeat inside my head:

"A scrotal rash on a gay man is syphilis until proven otherwise." I ordered a rapid test for syphilis and HIV, sent additional STI tests to the lab, and told Daniel that these two spots might be a sign of syphilis. He gestured toward the poster on the wall of the exam room, saying, "My spots don't look anything like that." The poster showed classic pictures of syphilis: an open sore on the penis, dark spots all over the torso, palms, and soles of the feet. True, Daniel's symptoms were subtle, but one tricky aspect of syphilis is how symptoms can look wildly different for different people.

When syphilis first works its way into a break in the skin of the genitals, mouth, or anus, it begins to multiply at the point of entry and then enters the lymph system and the bloodstream. Within days after being infected, *Treponema pallidum* are scattered throughout the body and can be found in almost any organ, from the genitals to the eyes and everything in between.[18]

Ten days to three months after having sex that exposes you, a painless ulcer (also called a *chancre*) appears, signaling the primary stage of infection. Usually the chancre is on the genitals, but it can also appear anywhere on the body that is used for sex. The chancre is moist in the center and firm around the outside, just like the cartilage at the end of your nose. While chancres on the penis are hard to miss, they can also form inside the vagina, the mouth, or the rectum, where they can easily go unnoticed.

Syphilis is an infection that thrives on denial. If Daniel had seen a chancre and convinced himself that he'd just caught his penis in his zipper, the chancre would have healed on its own and disappeared. He would have felt nothing, except perhaps relief that his penis was back to normal. Meanwhile, *Treponema pallidum* would quietly multiply all over his body.

If not caught and treated at the primary stage, six weeks to six months after having sex, the secondary stage begins. Classically, a rash erupts on the torso, eventually spreading to the hands and feet, and sometimes appears on the scrotum. There may be two spots or two hundred. It can look like ringworm, chicken pox, psoriasis, or almost any other skin condition under the sun. Most people will experience muscle aches, fatigue, and headache, but those can easily be mistaken for some other viral illness or the effects of stress. Occasionally, rarer signs of secondary syphilis like patchy hair loss or warty growths on the genitals might clue an astute clinician into the diagnosis.

At the secondary stage, antibody testing will almost always identify syphilis, but this depends on two factors: the patient needs to seek out medical attention, and the clinician needs to think about syphilis and test for it. In Daniel's case, the two spots on his scrotum (his only sign of a rash) and some enlarged lymph nodes were the only indication that he might have secondary syphilis. If there had been a rash on his torso, hands, or feet, it certainly wasn't visible now. His spots were found because he happened to walk into an STI clinic, where we examine every scrotum for the same reason that thieves rob banks—that's where the money is. If Daniel had walked into a typical urgent care clinic complaining of fatigue, his testicles might not be the first place a doctor would think to look.

If Daniel had chosen to ignore his fatigue, the signs of secondary syphilis would also disappear, entering a latent stage that could last for decades. At this stage, once again, all appears quiet on the body's surface. The patient's physical exam will be completely normal, but inside the organs, *Treponema pallidum* is still multiplying, albeit more slowly. At this latent stage, the only way to detect syphilis is with antibody testing.

If left unchecked, syphilis can cause destructive tumors of the skin and bones and weaken the walls or blood vessels of the heart, causing aneurysms that can burst and lead to sudden death. Before the advent of antibiotic therapy, about one-third to one-half of patients with untreated syphilis would suffer the complications of late-stage disease. The central nervous system can be affected at any stage, but it is a prime target during this late or tertiary stage, leading to progressive paralysis, blindness, deafness, even insanity.

In her book, *Pox*, writer Deborah Hayden postulates that tertiary syphilis was at the root of Vincent van Gogh's wild technicolor visions, Al Capone's murderous rages, and Adolf Hitler's erratic behavior. Much of this is speculative, based on historical accounts rather than formal antibody testing.[19] But even with antibody testing, the tertiary stage can be tricky to diagnose because the tests may lose positivity after decades of infection. Once the central nervous system is damaged by tertiary syphilis, treatment often can't reverse the damage that has already been done.

Daniel's syphilis and HIV rapid test results returned twenty minutes later. The syphilis test was positive, and the HIV test was negative. Because of the spots on his scrotum, it seemed that Daniel had already entered the secondary stage of the disease; the primary stage had gone by unnoticed. Treatment was straightforward: a single shot of long-acting penicillin would cure him within a week. Even after more than seventy years since penicillin's introduction, *Treponema pallidum* has not developed resistance to this simple antibiotic regimen.

After our clinic nurse gave him a shot of penicillin in his behind, Daniel sat down with Rebecca, one of the city's public health disease intervention specialists, who tried to help him gather the names of his partners over the past six months. He had been sexually active with at least seven men, including his new partner, but wasn't in touch with any of the others. He texted his current boyfriend to let him know that he'd need treatment. For the others, he just had profile names from a dating app. He disclosed the ones he could remember but was doubtful they could be found.

I sat and spoke with Daniel again, this time about HIV. His negative HIV test was a good sign, but syphilis was a red flag that he was having sex in high-risk networks. Syphilis and HIV are often referred to as *syndemics*—epidemics that are interrelated and affected by one another. A study by Preethi Pathela in New York City found that one in eighteen HIV-negative men would become HIV positive within a year of catching syphilis.[20] While Daniel had only contracted syphilis this time, perhaps next time, it would be worse.

I asked if he wanted to discuss other ways to prevent HIV, like taking daily pre-exposure prophylaxis (PrEP). Daniel was willing to learn more, but he wasn't ready to start anything new. He took some information and promised to think about it. Before he left, I reminded him that anal sex and oral sex were off the table for the next week, to allow enough time for the penicillin to get rid of the syphilis. "How about hand jobs?" he asked hopefully.

I smiled. "Yeah, sure. Go knock yourself out."

Daniel was lucky. He had been living with syphilis for months yet suffered no complications. A week after treatment, he would be completely back to normal. Unfortunately, this isn't always the case. When syphilis strikes during pregnancy, there is a race against the clock to treat the mother before the infection strikes the baby. If and when syphilis attacks a baby, it can often be a matter of life or death.

The Kids Are Not All Right

At the Community Regional Medical Center in Fresno, California, newborn Eliana was sleeping soundly in her incubator as Christian Faulkenberry-Miranda stopped by on her morning rounds. Eliana had been born almost a month early—at least that's what Faulkenberry and the other pediatricians had guessed. In truth, no one knew Eliana's exact gestational age, because her mother didn't have any idea how long she had been pregnant at the time of her delivery.

Eliana's mother had been living on the streets since the age of eighteen, when she ran away from a foster home and became addicted to crystal methamphetamine. When she became pregnant with Eliana at age twenty, she was so far gone with her addiction that she didn't seek any prenatal care. The first time she was seen by a doctor was when she presented to the emergency room in labor.

Without reliable information from Eliana's mother, Faulkenberry and her team pieced together their best estimate of Eliana's age based on her Ballard score, which accounted for neuromuscular tone, plus the appearance of body hair, skin, cartilage, and other physical features. Eliana was certainly on the thin side, with twiglike limbs and a protuberant belly, but Faulkenberry couldn't tell whether she was truly underweight or just missing the rolls of fat often found on full-term newborns.

Eliana's heart rate and blood pressure were stable, so Faulkenberry took a moment to examine the newborn's body more closely. Eliana had clear fluid running from her nose, but her breathing was unlabored, and her lungs sounded clear. As Faulkenberry pressed her fingers against Eliana's abdomen, she felt the edge of an enlarged liver, which was accounting for her belly's swollen appearance. There was an impressive rash on Eliana's torso, hands, and feet, which Faulkenberry described as a "blueberry muffin, like every rash you see on every exam you take on pediatric infectious diseases." The blueberry muffin rash is classically associated with rubella, but can appear because of other congenital infections, leukemia, and other blood disorders.

To determine what was going on, Eliana was going to need a battery of blood tests, x-rays to look for bone deformities, a spinal tap, and intravenous antibiotics. Although she appeared stable now, Faulkenberry decided to move her to the neonatal intensive care unit (NICU) just in case things took a turn for the worse. A few hours before Eliana was born, they'd tested her mother for rubella, syphilis, HIV, and hepatitis B, tests normally done at a first prenatal visit. The rapid HIV test was ready in less than an hour, so they knew it was negative before Eliana was born. But because Eliana was born on a Friday, the pediatricians would need to wait until Monday for the other test results to come back.

On Monday morning, Faulkenberry wasn't surprised when she logged in and found Eliana's and her mother's tests were positive for syphilis. The obstetricians had examined Eliana's mother thoroughly and found no physical signs of infection; during the pregnancy, she might have passed through the primary and secondary stages of syphilis and was now in a latent stage of infection. They'd never know how long Eliana had been exposed to syphilis before she was born, but now the constellation of Faulkenberry's physical findings made sense. Eliana's runny nose was not ordinary mucus, it was full of *Treponema pallidum* bacteria. And her liver was swollen from producing extra blood cells in response to the inflammation from the infection.

Eliana had been stable over the weekend, so by her fourth day of life, she was transferred out of the NICU to complete ten days of intravenous penicillin, the standard treatment for congenital syphilis. She began to recover quickly; the blueberry muffin rash was already fading by the time she was discharged. Eliana's mother agreed to go into drug treatment, and Child Protective Services allowed the baby to be discharged with her to stay with a relative. When Faulkenberry saw Eliana again at one month of age, her liver had returned to normal size, and she had put on weight and filled out her thin frame. From outside appearances, it was impossible to tell that she'd ever been ill.

Faulkenberry has become a de facto expert on cases like Eliana's, but certainly not by design. When she decided to specialize in pediatrics, congenital syphilis was a condition she had read about but never seen. She began medical school in 2000, the year when syphilis cases reached an all-time low in the United States. During four years of medical school and three years of pediatrics residency, she'd never seen a single case.

It wasn't until 2010 when Faulkenberry had her first babies on the pediatrics service who had been exposed to syphilis in utero. It gradually increased from a few cases per year to about one per month. By 2019, she was caring for newborns exposed to syphilis every week. Some of them were treated early enough to be born healthy. Others were like Eliana—ill babies who responded well to antibiotics. Then there were the babies who were stillborn or born with deformities.

It pained Faulkenberry to see this happening in her own hometown. She had grown up in Fresno during the 1980s and had witnessed its growth as an agricultural powerhouse of California's Central Valley.[21] Yet despite its agrarian wealth, one in four of the city's half million residents lives in poverty. According to the American Communities Survey in 2017, the median household income for a family of four was just under $49,000, compared to more than $96,000 in the city of San Francisco.[22]

The intersection of poverty and rurality has also provided fertile ground for crystal methamphetamine to thrive. Upon witnessing Fresno's drug problem firsthand, filmmaker Louis Theroux was compelled to film his addiction documentary there, calling Fresno *The City Addicted to Crystal Meth*.

Perhaps it was the crystal meth boom combined with the city's bare-bones public health infrastructure that caused congenital syphilis to spiral out of control. In 2011, there were only two cases of congenital syphilis in Fresno, similar to other neighboring counties in the Central Valley. By 2015, that number had increased to forty-two. That year the number of congenital cases in Fresno exceeded every other county in California, including Los Angeles County, which has a population ten times its size.

Transmission of congenital syphilis in utero is preventable if the mother can be treated with penicillin at least a month before delivery. But this was easier said than done, as Fresno's epidemic was concentrated among women like Eliana's mother, who were experiencing homelessness or struggling with mental illness, addiction, or incarceration. Cases like that of Eliana's were incredibly challenging to prevent; her mother had almost no contact with the health care system or social services prior to delivery.

There were also women who were tested for syphilis during pregnancy but didn't return for results and treatment. If the clinics couldn't reach the patient, then the public health department was supposed to step in to find patients and get them into medical care. But as cases of syphilis among both men and women began to increase, overwhelmed county public health workers began to fall behind on investigations. When cases were eventually investigated, sometimes the women could not be located or couldn't be treated soon enough to prevent transmission to the baby.

In March 2016, it was clear that things were going to get worse before they got better. Fresno County called for reinforcements. Disease intervention specialists, epidemiologists, and staff from the state health department and the CDC were dispatched to aid in the response. New systems were put in place to clear the backlog of cases, but the underlying social issues plaguing the county still exist. In 2017, the number of congenital syphilis cases in Fresno County rose to sixty before finally falling in 2018—there is still much work to be done.[23]

Meanwhile, what was happening in Fresno was happening

in pockets all over the United States, albeit in smaller numbers. As the CDC reviewed their surveillance data for 2018, it was clear that increases in congenital syphilis were outpacing other STIs. In October 2019, the CDC issued a press release to share the grim statistics. In the prior year, there were 1,306 cases of congenital syphilis in the United States, the highest number reported in twenty years. This was staggering compared to 66 cases of mother-to-child transmission of HIV, which the CDC had reported the same year.

"To protect every baby, we have to start by protecting every mother," Gail Bolan, director of CDC's Division of STD Prevention, said in the press release. "Early testing and prompt treatment to cure any infections are critical first steps, but too many women are falling through the cracks of the system. If we are going to reverse the resurgence of congenital syphilis, that has to change."

There are many cracks to fill, but the tools exist: testing (albeit flawed) and treatment for syphilis is inexpensive and effective. Women with risk factors or living in high-risk areas need to be tested multiple times during pregnancy to catch infections as early as possible. But just recommending more testing is not enough. Women with syphilis are increasingly reporting methamphetamine, heroin, or other injection drug use, and these are the most difficult women to find. In the western United States in 2018, CDC reported one-third of women with primary and secondary syphilis had used methamphetamine in the previous year.

In a letter to the STI prevention community in February 2019, Bolan acknowledged that the intersection of the drug and STI epidemics would require a coordinated public health response. She emphasized that health departments' STI programs would have to cooperate with needle exchange and substance-use treatment programs. The only way to break the cycle would be to adopt a holistic approach, providing multiple services to address patients' complex life challenges.

A coordinated response to both epidemics will certainly require more funding for health departments' STI and substance use prevention programs. There need to be enough public health disease intervention specialists to find the pregnant women who need testing and treatment and bring them to care. Still, there is no question that it must be done. The consequences of inaction—disabled or stillborn babies—are simply unbearable.

At least the resilience of youth is on our side. Pablo Sanchez, one of the nation's foremost experts on congenital syphilis, has a positive attitude about the potential outcomes. "If these babies are caught and treated early, they really do well." Despite his somewhat depressing choice of a specialty, he's seen enough success stories to know that things often turn out all right in the end.

I once saw a photo of one of his patients that he had cared for when she was an ill newborn—and now she was a preschool-age girl sitting in a white dress with red shoes, ribbons trailing from her head, grinning ear to ear. She'd made a complete recovery from congenital syphilis and blossomed into a healthy kid. He kept in touch with her over the years as she grew out of girlhood into adolescence. The last time Sanchez had spoken with her, she was about to graduate from college.

Early on in Sanchez's career, he actually cared for nineteen babies that had a similar situation to this patient's. They all had spinal taps demonstrating syphilis in their central nervous systems. Sanchez even took the trouble to confirm their diagnoses with rabbit infectivity testing (remember the testicles?). He knows that some are doing well as adults, but he wishes that he had kept closer tabs on all of them and tested their neurologic development to see how things turned out. He can only hope that they all recovered as well as the girl in the photograph.

As for baby Eliana, Christian Faulkenberry-Miranda has seen her periodically over the past four years. She's petite but has been growing normally. Her mom has been off methamphetamine for years, but she can't take back the exposure Eliana had while she was pregnant. Eliana's neurologic development may have suffered as a result. "She [Eliana] has some attention issues, but who knows. It's hard to know what was from the infection, what is genetic, and what was caused by drugs, or the environment." For the time being, Faulkenberry is patching up babies with syphilis as well as she can. "I know the treatment algorithm. I can handle this in my sleep. But it's so sad. This is something completely preventable."

Cases like Eliana's herald syphilis's troubling future, where babies suffer needlessly despite availability of effective therapy. But equally troubling is syphilis's ignoble past, one where antibiotic therapy was withheld from Black men with syphilis for decades in an infamous government research project—the Tuskegee syphilis study.

Originally entitled Untreated Syphilis in the Negro Male, this study was conducted by the U.S. Public Health Service (USPHS) under Surgeon General Thomas Parran to document the effects of untreated syphilis among 399 Black men in Macon County, Alabama. Initially meant to last six months, the study spanned from 1932 to 1972, extending for thirty years after the discovery of penicillin as an effective treatment for syphilis.[24]

Jean Heller of the Associated Press brought Tuskegee's abuses to light with a front-page story in *The New York Times* in 1972: SYPHILIS VICTIMS IN U.S. STUDY WENT UNTREATED FOR 40 YEARS. The government was left with substantial blood on its hands: twenty-eight of the men had already died from syphilis, one hundred died from related complications, forty of their wives had become infected, and nineteen of their children had been born with congenital syphilis.[25]

The public outcry after Heller's exposé led to sweeping policy changes to protect human research subjects that remain in place today. Still, the government's reputation had been greatly tarnished. Two decades later, Surgeon General Joycelyn Elders commented in *The Search for the Legacy of the USPHS Syphilis Study at Tuskegee:* "There can never be a greater scar or a more open wound for U.S. medicine or the U.S. Public Health Service (USPHS)."[26]

And for Black patients, the reforms were too little too late—their trust in scientific researchers had been irreversibly damaged. At least that had been the belief for decades, until a curious dentistry professor in New York began a quest to uncover the truth beneath our presumptions.

Searching for Tuskegee's Legacy

In the fall of 1993, Ralph Katz of New York University found himself on a twelve-hour train ride to the University of Virginia, headed to a conference entitled "The Tuskegee Legacy: Doing Bad in the Name of Good." At the time, Katz was directing a center on oral health among minorities and had been looking to enroll Black patients into his research studies. He thought the conference might inform his future recruitment efforts. To pass the time, Katz decided to reread *Bad Blood* by historian James Jones, the seminal book on the Tuskegee syphilis study he often assigned to his undergraduate and dental students.

At the conference, Katz listened as multiple speakers presented their impassioned views on the repercussions of Tuskegee—namely, how the recruitment of minorities into future research studies would be nearly impossible due to the study's legacy. After eight hours of lectures exploring the topic, Katz mused that "the legacy of the USPHS–Tuskegee study, however, appeared to be known more in the gut than in the head: everyone felt that it was true, but lacked any quantified research evidence on which to base those feelings."

When Katz returned to New York, he decided he would survey Blacks, Hispanics, and whites to see if they remembered the Tuskegee syphilis study and to evaluate how it might influence their decision to participate in research today. As Katz began to design his survey, he formed collaborations with social scientists, historians, and ethics experts with expertise in the Tuskegee study. He also joined forces with Reuben Warren, a Black dentist who was serving as the associate director for minority health at the CDC, and James Ferguson, one of the deans of Tuskegee University.

In January 1996, Katz's team and the outside experts met to review and critique the survey questions that would be used during Katz's study, now called the Tuskegee Legacy Project.[27] A subset of the group also developed a strategic plan to obtain an apology from the United States government to the Black community for the abuses of Tuskegee. Meanwhile, Katz began writing grants to secure additional funding. Some of his collaborators received angry calls when they heard about Katz's proposals. One caller asked Reuben Warren, "Have you forgotten that Katz is white?"

Katz understood. What business did a white Jew have studying the legacy of Tuskegee, a symbol of whites' racism and abuse of Black men? He realized that having himself at the helm of the Tuskegee Legacy Project might be, in his words, "unseemly." He also worried that he'd somehow be biased or that he'd misinterpret the data because he was simply too inexperienced. To mitigate this, he recruited leading behavioral scientists along with Black, Native American, and Hispanic/Latino policy leaders to serve on an advisory committee. They all agreed to help Katz's team interpret whatever the findings might be.

As Katz continued his work, Surgeon General David Satcher invited him to serve on the committee that issued a request for a presidential apology. In May 1997, he stood by as President Clinton delivered his apology to the remaining Tuskegee survivors in the East Room of the White House:

> To the survivors, to the wives and family members, the children and the grandchildren, I say what you know: No power on Earth can give you back the lives lost, the pain suffered, the years of internal torment and anguish. What was done cannot be undone. But we can end the silence. We can stop turning our heads away. We can look at you in the eye and finally say on behalf of the American people, what the United States government did was shameful, and I am sorry . . . And without remembering it, we cannot make amends and we cannot go forward.[28]

By 1999, Katz had enough funding to administer the Tuskegee Legacy Project Questionnaire in Hartford, Connecticut; San Antonio, Texas; and the cities of Birmingham and Tuskegee, Alabama. Of the 1,100 Black, Hispanic, and white adults that responded, less than a third said they were likely to participate in research, but that was true regardless of race. Blacks reported that they were more fearful of research than whites, but were *not* less likely to participate if offered the opportunity.

In 2003, Katz conducted another survey of more than 1,100 adults, this time in Baltimore, New York City, and San Juan,

Puerto Rico. This time he asked participants whether they'd ever "read about, or ever heard of any incidents or events related to medical studies or diseases that have ever affected your trust in medical research?"

He found that more than 89 percent of Blacks, whites, and Puerto Rican Hispanics couldn't name or identify the Tuskegee syphilis study. When he asked specifically about the study by name, about half of Blacks, a third of whites, and a quarter of Puerto Ricans had heard of it, but few (37 percent of Blacks, 27 percent of whites, and 9 percent of Hispanics) were able to provide any specific details. For those who offered details, myths about the study were almost as common as facts, including the beliefs that: 1) the study was conducted on soldiers, and 2) that the government deliberately infected the men with syphilis.[29,30]

Despite common wisdom that minorities wouldn't participate in research because of Tuskegee, Katz's data said otherwise. He concluded it was "unlikely that detailed knowledge of the Tuskegee Syphilis Study has any current widespread influence on the willingness of minorities to participate in biomedical research. Caution should be applied before assuming what community leaders 'know and are aware of' is equally 'well known' within their community constituencies."

There's no question that Tuskegee was and remains a powerful metaphor for racism and exploitation of human research subjects, but Katz felt that "Blacks would not have needed the Tuskegee Syphilis Study to be trepidatious about entering into any white male dominated activity (e.g., research) in this country." Aside from Tuskegee, Blacks had survived four hundred years of abuses contributing to intergenerational trauma and mistrust of the white establishment: slavery, Jim Crow, police brutality, mass incarceration. In Katz's opinion, Tuskegee's perceived legacy should not be used by researchers as an excuse to forgo recruiting minority participants in research. Researchers would need to work harder to establish trust in minority communities, but recruitment could and should be done.

While most of the public doesn't remember Tuskegee's past, there were some for whom the memories of the study would never fade. Among the Tuskegee survivors, the last participant died in January 2004, the last widow in January 2009. According to the CDC, twelve remaining descendants of the Tuskegee syphilis study are still receiving health and medical benefits paid from the federal government—a small reparation for the sins of the past.[31] Universities and research organizations haven't forgotten either. Because of Tuskegee, there are now institutional review boards to protect the rights of study participants and ensure that they provide informed consent. With the current protections we have in place today, we will not be doomed to repeat the mistakes of our past.

Unfortunately, Tuskegee was not the only STI research being conducted by the U.S. government under the public's radar during this period. In 1946, one thousand miles to the south in Guatemala, the U.S. Public Health Service was at work on another shameful set of experiments. This time, several myths of Tuskegee would be realized as truths: the government deliberately infected soldiers with STIs, as well as prisoners, sex workers, and the mentally ill. It is an astonishing tale of unethical conduct, whose entire story is still yet untold.

8

The Path of Least Resistance

Sailors, Sex Parties, and Gonorrhea Superbugs

Ethically Impossible

As the United States and its allies fought bravely during World War II, our government feared the Axis powers had a weapon that could destroy our forces from within. This was no ordinary bomb or missile—it was rampant venereal disease (VD) that posed the greatest threat to our country's readiness for battle. These were the claims splashed across government propaganda posters, promoting use of prophylaxis and avoidance of prostitution: "You can't beat the Axis if you get VD," "Our carelessness, their secret weapon," "You Can't Slap a Jap with the Clap."[1]

The government's concerns were certainly justified. At the start of World War II, Dr. Joseph Earle Moore, chairman of the National Research Council's VD subcommittee, predicted 350,000 gonorrhea infections in the military would "account for 7,000,000 lost man days per year," the equivalent of putting the strength of ten aircraft carriers out of action. Soldiers with acute gonorrhea often developed painful urination, discharge, and pain or swelling in their testicles, rendering them unable to fight. The United States had already suffered such venereal casualties of war during World War I, when the army lost almost 7 million person-days and discharged more than 10,000 men due to syphilis and gonorrhea. Only the great influenza pandemic of 1918 caused a larger loss of duty.[2]

To prevent STIs during World War II, the military was recommending the same postcoital prophylaxis regimen they'd used for more than thirty years. First, soldiers were asked to urinate, then wash their penises with soap and water. Silver proteinate ointment was then injected into the urethra, followed by a slathering of mercury-based calomel ointment over the penis and pubic region. The penis would be wrapped in wax paper to catch any leakage. The wax paper crinkled and crunched, making it obvious to fellow soldiers when someone had received prophylaxis. At night in the barracks, a chorus of crackling penises could be heard as soldiers rolled over the wax paper in their sleep.

Because the prophylaxis regimen was embarrassing and messy, it wasn't well liked among the enlisted men. Still, it might have been fine if it prevented gonorrhea. But after more than three decades of use, military scientists still weren't sure whether this prophylaxis regimen was effective. They decided to embark on a new study of gonorrhea prophylaxis regimens, testing two different treatments: 1) sulfa drugs taken by mouth before exposure, compared to 2) ointments injected into the urethra after exposure.

In 1943, the U.S. Public Health Service (USPHS) conducted these gonorrhea prophylaxis experiments at the federal penitentiary in Terre Haute, Indiana, with prisoners serving as volunteers. About five months in, the project's leader, Dr. John Mahoney, raised serious concerns to the National Research

Council (NRC). The researchers had tried depositing strains of gonorrhea from the laboratory at different concentrations into the ends of the prisoners' penises, but they could not consistently infect the prisoners through artificial exposure. By June of 1944, Mahoney reported that they would be calling it quits.

The inability to induce gonorrhea in the prisoners in Terre Haute left the USPHS researchers without an answer to their question of what worked as effective prophylaxis. What they hadn't tried was directly inoculating the penis with discharge taken from an infected woman or introducing gonorrhea by a more natural and potentially effective experiment—sexual intercourse with an infected partner. Sex might be a more reliable means of transmitting gonorrhea, but how could they conduct such an experiment?

At the USPHS Venereal Disease Research Laboratory (VDRL) on Staten Island, physician Juan Funes, a visiting Guatemalan research fellow, suggested to Mahoney that Guatemala could serve as a potential location for the study. At the time, he served as director of the Guatemalan Venereal Disease Control Department, and felt it was feasible to do a study with Guatemalan prisoners similar to that in Terre Haute. In Guatemala, it would be legal to bring sex workers with gonorrhea into the prisons to have sex with the prisoners, creating a natural exposure scenario likely to result in infection.

The NRC agreed to fund the study, and John Cutler, a young physician who had assisted with the prisoner studies in Terre Haute, and Juan Funes were chosen to spearhead the investigations. The plan for the Guatemala studies was to test a new compound, orvus-mapharsen, as postexposure prophylaxis for gonorrhea along with syphilis and chancroid (another bacterial STI causing genital ulcers).

In 1946, Cutler and colleagues went to Guatemala to set up

a laboratory and field site for the study. Commercial sex workers were recruited so they could be intentionally infected with the gonorrhea that had been cultivated in Cutler's laboratory. Funes also referred sex workers who were already infected from the country's Hospital de Profilaxis (Venereal Disease and Sexual Prophylaxis Hospital). The women had intercourse with Guatemalan soldiers or prisoners, who were examined immediately afterward and followed over time to look for signs of infection. For some of the experiments, workers were compensated with twenty-five dollars for their efforts.

On occasion, both the sex workers and the research participants were plied with alcohol, with the rationale that drinking might lower resistance to infection for men. The motivation for liquoring up the women was not as clear. In reviewing Cutler's papers from Gonorrheal Experiment #4 in the National Archives, it seems the relentless pace of intercourse might have warranted some liquid courage: one experiment describes a sex worker servicing eight soldiers in under ninety minutes. Cutler was puzzled that Guatemalan sex workers weren't more enthusiastic toward his work: "Contrary to what might be expected, it proved extremely difficult to obtain prostitutes willing to serve under experimental conditions."[3] Even more frustrating for Cutler, only 10 percent of the sexual contacts resulted in successful transmission of gonorrhea.

By 1948, Cutler had completed 32 gonorrhea intentional exposure experiments, including 582 participants: soldiers, prisoners, sex workers, and a few dozen psychiatric patients.[4] None of the research subjects were taken through the process of informed consent. Cutler then prepared final reports on the studies in Guatemala and sent them to the director of the VDRL between 1952 and 1955. The reports were all marked as SECRET-CONFIDENTIAL.

Notably, the data from the intentional exposure studies were never published by Cutler or any of his colleagues. Cutler wasn't ashamed of the subject matter; at the time, these types of studies were not taboo. In 1956, he coauthored a comprehensive review of other intentional exposure studies without making mention of his Guatemala work. Perhaps it was because the researchers didn't successfully infect enough subjects to get useful data on prophylaxis measures. Or maybe the military or U.S. government somehow prevented Cutler from publishing his findings. No one knows. The reasons the experiments ended and the secrecy of the data remain a mystery.

John Cutler, John Mahoney, Juan Funes, and the other key investigators all went on to have illustrious careers, their success unscathed by their actions in Guatemala. And as if the Guatemala experiments weren't morally hazardous enough, Cutler would continue his work in the USPHS as the lead researcher for the Tuskegee syphilis study from 1951 to 1954. Unlike Guatemala, his work in Tuskegee was not conducted in secret—the scientific community was aware of it through multiple publications prior to its public exposé in 1972. Unfortunately, this awareness did nothing to halt the study's conduct.

The Guatemala experiments might have stayed secret had it not been for Susan Reverby, a history professor from Wellesley College. In 2008, while writing her book *Examining Tuskegee,* Reverby began digging through Cutler's papers in the library archives at the University of Pittsburgh, where Cutler had once been a dean at the Graduate School of Public Health. Instead of new documents about Tuskegee, she came across more than twelve thousand pages of documentation from Guatemala entitled: "Experimental Studies on Human Inoculation in Syphilis, Gonorrhea, and Chancroid."[5] It consisted of correspondence be-

tween Cutler and his colleagues, laboratory and experimental reports, and graphic photographs.

Reverby was shocked. Cutler had donated these documents
to the university's archives in 1990, and they had been sitting
there for almost twenty years. Anyone could have found them,
and yet no one had. She combed through Cutler's papers and
wrote up a summary of her findings. She presented these to a
modest audience of twenty historians at the American Association for the History of Medicine meeting in May of 2010. She
planned to publish an article of her findings in early 2011 in a
policy history journal.

Her work might have remained within a small niche of academia, except that Reverby showed a draft of her unpublished
paper to David Sencer. Sencer was a former director of the CDC
whom she had interviewed for her book on Tuskegee. He was
disturbed by her findings and passed her paper to the leadership of the CDC, who sent one of their experts to the University
of Pittsburgh to verify Reverby's discovery. Their expert's report
and Reverby's unpublished article then wound its way up the
chain of command, finally reaching the White House during the
summer of 2010.

On October 1, 2010, President Barack Obama called President Álvaro Colom of Guatemala to apologize, and requested a
formal investigation. Subsequently, the Presidential Commission
for the Study of Bioethical Issues published a report in September 2011, titled "Ethically Impossible," which summarized how
Cutler managed to exploit almost every group designated as a
vulnerable population in federally funded research today: prisoners, military members, children, the mentally ill, the econom-
ically or educationally disadvantaged.[6]

From our current perspective, the abuses in Guatemala are

clearly egregious. But seen through a historical lens, was this usual conduct for the time? Based on the correspondence between investigators and colleagues throughout the study, it's clear that they and perhaps the surgeon general knew that they were on shaky moral ground:

February 1947, G. Robert Coatney (malariologist, National Research Council) to Cutler:

> As you well know, [Dr. Parran] is very much interested in the project and a merry twinkle came into his eye when he said "you know, we couldn't do such an experiment in this country." [Thomas Parran was surgeon general in 1947.]

June 1947, Cutler to Mahoney:

> It is imperative that the least possible be known and said about this project, for a few words to the wrong person here or even at home might wreck it or parts of it. We have found out that there has been more talk here then *[sic]* we like with knowledge of the work turning up in queer places.

April 1948, Richard Arnold (penicillin researcher) to Cutler:

> I am a bit, in fact more than a bit, leery of the experiment with the insane people. They cannot give consent, do not know what is going on, and if some goody organization got wind of the work they would raise a lot of smoke. I think the soldiers would be best or the prisoners where they can give consent.[7]

Unlike with Tuskegee, the U.S. government has not provided compensation or medical benefits to Guatemalan survivors or their families. Several Guatemalans brought a lawsuit against the U.S. government in 2012, but it was dismissed on the grounds of sovereign immunity under the Federal Tort Claims Act, which "bars claims based on any injury suffered in a foreign country, regardless of where the tortious act or omission occurred."[8,9]

In 2015, Guatemalan victims and their families launched a civil suit against Johns Hopkins University, Bristol Myers Squibb, and the Rockefeller Foundation, claiming that the three defendants' employees played crucial roles in the Guatemala experiments. The defendants filed a motion for dismissal. In January 2019, a federal judge denied their motion, allowing the *Estate of Arturo Giron Alvarez v. Johns Hopkins University et al.* to have its day in court.[10] In a piece for *Slate*, journalist Sushma Subramanian interviewed one of the plaintiffs, ninety-one-year-old Frederico Ramos, who was fatalistic about the outcome. "When I found out about the Americans, I put myself in the hands of God. I prayed to get some compensation for it." But ultimately, he realized, "God will decide."

After much pain and suffering inflicted on patients in Guatemala, the experiments did not result in discovery of the perfect prophylaxis to prevent gonorrhea. Instead, it was the introduction of penicillin in the 1940s that would revolutionize both the treatment and prophylaxis of gonorrhea during wartime. During the Korean War, penicillin prophylaxis was highly effective when given to troops before liberty periods, almost eliminating new cases of gonorrhea in two units that were studied. According to Major Mark Rasnake of the U.S. Air Force, this prophylaxis was then authorized for general use.[2] Penicillin was the weapon that ensured victory in the military's battle against venereal disease.

Unfortunately, the victory would turn out to be short-lived. By the Vietnam era, gonorrhea began requiring increasing doses of penicillin to achieve a cure. During wartime, rumors began circulating about soldiers with incurable gonorrhea in the Far East, but no one had verified that they were true (these were the rumors feared by Mike the biker in chapter 6). But military scientists had never transported gonorrhea specimens back from Asia and tested them to prove that they couldn't be cured with penicillin. That would all change in the spring of 1964, when Lieutenant King Holmes headed from Tennessee toward the Pacific to report for duty, not knowing that the navy would sail him straight into gonorrhea's awaiting arms.

Hello, Sailor

A decade before he would start building an STI research empire in Seattle, King Holmes started his career as an internal medicine resident at Vanderbilt University Medical Center in Nashville, Tennessee. In May of 1964, he received a call from Captain Herbert Stoecklein telling him that he was going to be drafted. Holmes listened politely as Stoecklein explained the details of his deployment.

"Well, boy, we're going to send you to China Lake."

"China Lake? Sir, that's fabulous, I didn't know you had bases in China."

"Well, boy, it's not in China. It's in Barstow, California."

"Barstow? That's the Mojave Desert."

The wheels turned rapidly in Holmes's brain. He had already developed a penchant for infectious disease research and was hoping for something more academic than a military base in the middle of the desert. He decided to try negotiating. "Sir, if you would either send me to Japan to one of your research labs or to Hawaii to the Preventive Medicine Unit, I would sign up for a third year [of deployment]."

There was a long pause as Stoecklein mulled over Holmes's proposal, before relenting. "Well, okay. If we do this, I'd be watching you. If you didn't sign up for the third year I'm shipping your ass to Vietnam, family or no family."

This is how Holmes had the good fortune of spending the Vietnam War as a lieutenant in the navy's Preventive Medicine Unit in Oahu, Hawaii. A few months into his stay, Holmes ran into David Johnson, an epidemiologist for the navy's Seventh Fleet. He told Johnson how much he'd like to have his job someday. Two days later, he received a call from Johnson asking, "Were you serious?" Johnson had just been promoted and needed someone to replace him. A few weeks later, Holmes had arranged for someone to cover his clinical duties; he had just become the Seventh Fleet's new epidemiologist.

By now, Holmes had heard the rumors about sailors stationed in the Pacific with incurable gonorrhea. His first task as the Seventh Fleet's epidemiologist was to figure out what was going on. "We were going to have to figure out where the men were being seen, and what they actually had." It was possible that some of these cases were "incurable" simply because they were infections other than gonorrhea. He would need to obtain specimens from patients with gonorrhea, grow the bacteria in a culture, and test them to see if they were resistant to antibiotics. There was one small issue: Holmes didn't know how to do any of this. He'd need more training than he currently had.

Holmes decided to pursue a Ph.D. in microbiology at the University of Hawaii, but soon had to interrupt his studies as more reports of incurable gonorrhea trickled in from the Seventh Fleet in the Pacific Islands—instead, setting sail aboard the USS *Enterprise* for the Philippines. He started conducting interviews with the sailors on the ship, and once they arrived in port, he tagged along with them to the town of Olongapo in Central

Luzon. There, Holmes met the mayor and the leadership of the local nightclub and bar owners' association, informing them he was there to help control gonorrhea—an idea that was met with great enthusiasm.

Among Olongapo's population of 107,000, Holmes estimated that there were almost 5,000 sex workers, each of whom needed to be examined for gonorrhea at the local public clinic.[11] The workers would queue up to hop one by one onto the exam table. The examiner would place a speculum briefly into the vagina, sweeping a cotton swab against the vaginal walls. The swab would be handed to a technician, who would stain it with methylene blue dye, and then review the slide under the microscope. The whole process took about two minutes. There was one highly efficient female clinician who did the exams, conducting hundreds in a single day.

If it looked like one of the women had gonorrhea, she would be treated with an injection of penicillin, the standard treatment at the time. It all sounded well and good at first, but as Holmes dug deeper, he discovered two glaring issues. First, the treatment given to the sex workers was a formulation of penicillin called *benzathine*, which stays in low concentrations in the bloodstream for weeks. While it is an excellent treatment for syphilis, these low, steady concentrations of penicillin did not reliably kill gonorrhea. In fact, it created the perfect setup for gonorrhea to develop penicillin resistance. Then Holmes noticed that the clinician had no autoclave machine to sterilize her equipment. She had used the same infected speculum on hundreds of consecutive women, undoubtedly spreading gonorrhea among the sex workers in the process.

Luckily, both issues were easily fixed. Holmes recommended switching the treatment to procaine penicillin, which would quickly and effectively kill gonorrhea. His admiral was able to bring him an autoclave machine, which he donated to the local clinic, along with hundreds of metal specula for pelvic exams that could be sterilized and reused.

While Holmes had determined some factors that might have led to antibiotic resistance and the rapid spread of gonorrhea, he hadn't gotten his hands on any *Neisseria gonorrhoeae* to culture and test in the laboratory. On a subsequent trip, he collected specimens from sex workers and sailors with gonorrhea and placed them inside of a liquid nitrogen tank that he'd brought along on the journey. The -320°F temperature inside the tank put the bacteria into a state of suspended animation; Holmes planned to thaw and revive the bacteria once he returned to the United States. These specimens would be the key to understanding why certain cases of gonorrhea in the Pacific were so difficult to cure.

First, Holmes needed to get his nitrogen tank back to Hawaii, which turned out to be easier said than done. He would have to make his way to Clark Air Base, first crossing Manila Bay to reach the capital city, and then traveling forty miles northwest to the base. Once in Manila, the forty-mile journey to the air base took Holmes two days, requiring a combination of buses and wooden banca boats. As he approached the base, he passed a bus full of dead bodies, victims of a gas explosion aboard the bus. Holmes shuddered as he drove past and tightened his grip on his tank, grateful that both he and his gonorrhea had survived the journey.

At the base, the air force would not allow the tank that he'd brought all the way from the USS *Enterprise* onto their planes. They flatly refused to transport him and his cargo back to Hawaii. There was no way that Holmes was going to abandon his samples, which he referred to as "liquid gold" and "my precious gonorrhea," à la Gollum from *The Lord of the Rings*. He let everyone else in his group go back to the States without him, while he remained behind: "I am not leaving it," he insisted.

Holmes feared he'd be stuck, but the next day, a National Guard unit from Minnesota arrived at the air base. Holmes, a native Minnesotan, decided to work his hometown connections, getting friendly with the pilots and explaining his plight. The National Guard agreed to take him back to Hawaii on their transport plane. Holmes sat in the middle of the plane next to his liquid nitrogen tank, which had been tied down away from the cockpit by one of the guardsmen.

As they became airborne and started gaining altitude, fumes began emanating from his tank. The change in cabin pressure had caused his liquid nitrogen to become volatile and leak. One of the guardsmen serving as flight crew walked out of the cockpit at that moment, took one look at the steaming tank, and jumped back into the cockpit, slamming the door.

Moments later, he reopened the door and walked toward Holmes, saying, "We're going to have to dump this into the ocean. This is very dangerous."

Holmes had to think fast. "These are dangerous bacteria in this tank and we can't just dump it in the water, it could kill people." (Not true, but clever.) "I'll hold it so it won't tip over. We'll be all right."

Holmes gripped the tank between his legs for the rest of the flight. The National Guard's plane dropped him off in Hawaii, likely glad to be rid of him and his cargo. He rushed back to his laboratory and thawed his frozen *Neisseria gonorrhoeae* samples; most had survived the journey quite well. As he tested them, he found that some of the isolates had indeed developed high levels of resistance to penicillin. He could now verify the rumors that he'd heard about incurable gonorrhea; these sailors and sex workers would need something other than penicillin to cure them.

These were some of the world's first studies of penicillin resistance in *Neisseria gonorrhoeae,* but they certainly wouldn't be the last. Holmes didn't realize it at the time, but Hawaii was and still remains a hot spot for antibiotic-resistant gonorrhea. To respond, the state has developed a robust gonorrhea surveillance system—today, Hawaii tests a greater proportion of its gonorrhea cases for antibiotic resistance than any other U.S. state.

Thus, in April of 2016, the state laboratory was well equipped to detect a cluster of antibiotic-resistant gonorrhea cases, strains with resistance patterns never seen before in the United States. During the investigation into this cluster of cases, identifying and testing the gonorrhea was the easy part. But where did it come from, and where might it have spread? Solving that mystery would require detective work from the best of the health department's Hawaii Five-0, the disease intervention specialists.

No Happy Ending

According to the Hawaii Tourism Authority, almost ten million visitors came to the Aloha State in 2019, a number that has grown every year for the past seven years.[12] And who can blame them? Hawaii is full of gorgeous beaches, rainbows, waterfalls, and that famous aloha spirit—a feeling that is evoked by the warmth of the people and the beautiful scenery of the islands.

I came across a tourism brochure that tried to put this feeling into words: "When host and guest share in a genuine cultural exchange, an authentic friendship and respect is born. This is the 'Aloha Spirit.'" The brochure omitted one important fact: more often in Hawaii than in any other U.S. state, these cultural exchanges (particularly the naked ones) don't just generate Aloha Spirit—they result in cases of antibiotic-resistant gonorrhea.

Hawaii has served as our nation's portal of entry for antibiotic-resistant gonorrhea to the continental United States for the past forty years. Historically, the spread of antibiotic resistance in gonorrhea follows a recurring pattern, with cases appearing first in Japan and Southeast Asia, then coming to Hawaii before spreading eastward across the continental United States. Exactly why it tends to spread this way is unclear.

The first cases of penicillin resistance in the United States first appeared in Hawaii in 1976, followed by tetracycline in the '80s, and then ciprofloxacin in the '90s. After the first resistant cases appear in Hawaii, it takes several years for those strains of gonorrhea to spread sexually across the United States, but when they do, we effectively lose the ability to use that class of antibiotic for the treatment of gonorrhea.

In April and May of 2016, while conducting their usual surveillance activities, the state laboratory discovered eight isolates of *Neisseria gonorrhoeae* from seven different patients that were resistant to multiple antibiotics.[13] This ordinarily would not have raised any alarms, as the CDC found half the gonorrhea they tested in 2016 was resistant to at least one antibiotic. But these cases were resistant to four antibiotics: penicillin, tetracycline, ciprofloxacin, and azithromycin. Five years prior, in 2011, Hawaii had identified gonorrhea from a single patient with isolated azithromycin resistance—but these new cases were even worse. Not only were they resistant to four antibiotics, they also showed reduced susceptibility to ceftriaxone, the CDC's only remaining recommended treatment for gonorrhea.

The laboratory needed to be sure of what they had before raising any alarms. The state of Hawaii reached out to both the

CDC in Atlanta and the University of Washington Neisseria Reference Laboratory. They created subcultures of each of the eight isolates and sent them to both laboratories to confirm what they had found. In their own labs, they began sequencing the entire genome of the bacteria to see if the gonorrhea strains from the patients were genetically related to each other.

Meanwhile, the state's disease intervention specialists (DISs) were dispatched to find the patients and try to see if they could find any common links among them. All seven patients were located and interviewed: six men and one woman (none of whom had slept with each other). Three of the men had slept with sex workers, and two of the men named the same sex worker at the same massage parlor. Was this erotic masseuse the patient zero who set off this cluster of cases?

DIS staff headed to each of the massage parlors named by the patients. One establishment would not let them in the door. The second massage parlor allowed them inside, where they actually found the masseuse shared by two of the patients. She agreed to come into the health department's STI clinic to be tested.

To everyone's surprise, her tests were negative for gonorrhea. Alan Katz, the former director of the state's STD/AIDS Prevention Program, had some theories as to why this might be: "There was some concern that maybe the woman who ran the massage parlor had a private physician or had a supply of medications that they used to treat the workers [who might be suspected of having STIs]." Illicit massage parlors, particularly those trafficking women from Asia, have an incentive to keep workers out of sight of traditional medical clinics.

When resistance testing results returned from the CDC and the University of Washington, they verified what had been found by the Hawaii State Department of Health. The strains were resistant to four antibiotics, including azithromycin. The bacteria had reduced susceptibility to the drug ceftriaxone but were not completely resistant. Each of the patients was able to successfully clear their infection after receiving treatment with that antibiotic.

Although a patient zero was never identified, the sexual links between the patients and sex workers, plus the genetic sequencing tests, convinced the state epidemiologists that the isolates were closely related. The Hawaii Department of Health worried about this new strain of super gonorrhea that had already been transmitted to seven people. Would there be dozens more cases in the months to come?

Luckily, the worst-case scenario did not materialize. As of March 2020, there were no similar cases to the cluster observed in 2016. The Hawaiian strain somehow died out before being spread widely through the community. But we are certainly not out of the woods. Between 2017 and 2019, multiple cases of drug-resistant gonorrhea appeared throughout the world, including some cases that were resistant to ceftriaxone, the last remaining drug recommended for treatment. These cases appeared in Canada, Australia, and England; two of the patients required hospitalization and multiple days of intravenous antibiotics.

How can science and public health respond to this threat? A few new drugs are in development, but the antibiotic pipeline takes years to produce a new drug, and one may not be ready before an outbreak emerges. Testing for resistance could help, allowing us to use older antibiotics if a strain of gonorrhea happens to be sensitive to them. Since 1986, the CDC has been monitoring isolates of gonorrhea across the United States for antibiotic resistance, but test results were previously for surveillance only, not available in time to help guide treatment—and testing was only performed on a select number of men with infections of the penis.

Knowing that a wave of antibiotic-resistant gonorrhea will eventually reach our shores, the CDC is funding cities and states across the United States to ramp up surveillance: more antibiotic susceptibility testing, faster turnaround for results, rapid contact tracing of partners to nip outbreaks in the bud. Perhaps one of the most important surveillance activities is examining gonococcal infections outside the penis, focusing on those in the cervix, the rectum, and the throat.

When it comes to gonorrhea, the throat deserves special attention. At this moment, there could be unprotected sex going on inside your throat that you aren't aware of. I'm not talking about sex between you and another person (if that *is* happening, get your priorities straight and put this book down). I'm speaking of bacterial sex, the exchange of genes between *Neisseria gonorrhoeae* and other harmless bacteria that reside in your throat on a normal basis. And why should you care if your tonsils are the site of this bacterial hanky-panky? Open wide, say, "Aaah," and we'll see how and why antibiotic-resistant gonorrhea may be lurking inside.

Deep Throat

One Friday morning in June of 2018, I opened my work email to see a message from Yvonne Piper, a nurse practitioner who I work with at the STI clinic. The subject line read, "Sex party pharyngeal gonorrhea." At that moment, I had two thoughts: 1) I love my job, and 2) my spam filter had given up trying to protect me from lewd content (in my line of work, it's hopeless anyway). I continued reading Piper's email, which contained a list of instructions preparing the clinicians for a potential influx of patients seeking testing for pharyngeal gonorrhea (GC):

Recently we have had requests from cis [gender] women for pharyngeal GC testing. These women have self-identified as members of a sex party group on Facebook that has ~400 members. Some member of this group reportedly has gonorrhea. So far we have seen 4 people who are in this group; patients have let us know during the clinical visit that they are members of said group.

This is a pansexual play party group so expect you could have people of any gender coming in with this request. Make sure you take a good sexual history and only screen people who are performing fellatio—we are not offering this screening to people who report performing cunnilingus or analingus.

Piper has nothing against cunnilingus or analingus, but those activities are not as efficient at getting gonorrhea to the back of the throat. The clinic had to prioritize who to test among the four hundred members of the sex party group, and fellatio can deposit gonorrhea right on the tonsils, where it can happily establish an infection.

Despite the size of the group, we weren't expecting a horde of pansexuals with sore throats beating down our door. When *Neisseria gonorrhoeae* infects the throat, it rarely produces symptoms, enabling it to avoid detection on the part of its host. In most cases, the throat serves as a silent reservoir of infection; gonorrhea can only be detected there if someone thinks to look for it. Occasionally, it causes a sore throat and fever, mimicking a case of strep throat, but this is the exception, not the rule.

The silent infection is a clever survival strategy. To ensure propagation of the species, *Neisseria gonorrhoeae* must continue to be sexually transmitted to others. To do this, the bacteria needs to be hardy enough to withstand the environment of the throat, rectum, or genitals, and avoid being killed off by the host's immune system. It needs to stay silent or cause symptoms mild enough not to arouse the host's attention. If the host develops a raging case of tonsillitis or discharge, they are apt to throw antibiotics at it, dashing *Neisseria gonorrhoeae*'s chances of transmission.

Here's another bacterial survival strategy: hide somewhere where it's hard to kill you. Eradicating gonorrhea from the throat is more difficult than ridding it from other sites in the body, though the exact reasons for this are still unclear. Perhaps the immune response in the throat differs from that of the genitals or rectum, making it less likely that one will clear the infection at that site. It may be that antibiotics don't reach high enough concentrations for long enough to successfully clear infection from the throat.

Once inside the throat, gonorrhea can sit for weeks or months hanging out with its cousins, other species of *Neisseria* that have names reminiscent of hip-hop artists: *flava, subflava, sicca*. These are permanent residents of the mouth and aren't thought to cause any harm. Sometimes *Neisseria meningitidis* can also be present, a strain that can cause severe infections such as meningitis or also reside quietly without causing symptoms.

As they coexist together in the throat, the various *Neisseria* species exchange or donate pieces of DNA to each other, resulting in genes that are a combination or mosaic of the two different species. The bacteria with its new mosaic genes will multiply, be passed to someone's genitals through oral sex, and then perhaps to someone else's throat, where the process can occur again.

Over time, mutations to the chromosomes of *Neisseria gonorrhoeae* gradually morph the bacteria from its original state to one that is antibiotic resistant. It is a slow evolutionary process that happens over years, with multiple genetic changes required before antibiotic resistance develops.[14] For that reason, the development of resistance to penicillin was almost forty years in the making.

At some point, however, gonorrhea found a shortcut to antibiotic resistance by having bacterial sex or conjugating with another type of bacteria in the throat, most likely a species of *Haemophilus*. Scientists think that *Haemophilus* transferred over a plasmid, a circular piece of DNA that can replicate independently from the rest of the bacterial chromosomes—and that is capable of carrying multiple antibiotic resistance genes. Never mind having to go through years of evolution. One act of conjugation involving a plasmid, and voilà—instant antibiotic resistance.[15]

With two mechanisms for developing antibiotic resistance, it's no wonder that gonorrhea has managed to develop resistance to every antibiotic ever used to treat it. And now that we are down to one antibiotic (ceftriaxone) left to treat gonorrhea, what sounds like a bad B movie title is not so far from reality: *Attack of the Sexually Transmitted Superbugs*.

Indeed, mutant strains of pharyngeal and genital gonorrhea with resistance to ceftriaxone have reared their heads in the UK and other parts of Europe, Japan, and Australia from 2011 to the present day.[16] So far, there have been scattered cases here and there, though none has taken off and spread widely through the population. How could this be?

If this were indeed a B movie, the mutant bacterium would grow ever more monstrous, then become unstable and self-destruct. This hasn't really happened, but it is possible that these mutant strains of gonorrhea have reduced fitness, a concept related to a bacteria's ability to replicate and survive. If mutant antibiotic resistant strains are less fit than their predecessors, they may die out before they can replicate and be transmitted, one saving grace to stave off a pandemic of super gonorrhea.

If throat infections play a key role in antibiotic resistance, perhaps preventing them could help public health's gonorrhea response efforts. Using condoms for fellatio could likely help reduce transmission to the throat—that is, if anyone actually used them. I have asked patients whether they used condoms when performing fellatio, and people look at me like I am crazy. It simply isn't done, at least not by the majority of the sexually active public. According to a study by Melissa Habel of the CDC, more than 75 percent of heterosexual adults reported giving or receiving oral sex, but only 6–7 percent used a condom the last time they did.[17]

Some clever Aussies have decided to try a different tack. Researcher Eric Chow has spent years trying to figure out why men who have sex with men have such high rates of gonorrhea. Along the way, he's found gonorrhea in the saliva of men who also had gonorrhea in their throats, leading him to believe that saliva exchange during oral sex, kissing, or using saliva as lube during sex might be playing a role (which would also be true for women too).

With both the throat and saliva as potential culprits, Chow decided to throw some Listerine at the problem to see if it could help. And why not? Listerine was invented in 1879 as a surgical antiseptic, but later marketed as a floor cleaner, a vaginal douche, a cure for dandruff, and yes, even a treatment for gonorrhea (not sure exactly how it was applied).

In 2015, Chow conducted a small randomized controlled trial that included fifty-eight men with pharyngeal gonorrhea who had sex with men. He found that about half of men with gonorrhea who gargled with mouthwash for sixty seconds still had a positive throat culture, compared to 84 percent who just gargled with saline. Of note, this was serious gargling to the back of the throat, not a simple swish and spit.

Chow decided to go bigger and better with his idea, this time enrolling 530 Australian men who have sex with men in a study to see if mouthwash could actually prevent infections. The OMEGA Study (Oral Mouthwash used to Eradicate GonorrhoeA) involved daily gargling with two competing mouthwashes to see if either might reduce infections with pharyngeal gonorrhea.[18] If a simple mouthwash could prevent gonorrhea infections of the throat, it could be a huge boon for gonorrhea control worldwide. Not to mention the pleasant minty breath that would pervade bars and nightclubs the world over.

While the final study results have yet to be released, in a 2019 interview for the Australian magazine *Emen8*, Chow warned people not to go crazy with mouthwash or stick it into other parts of their bodies. "I can definitely say you shouldn't apply mouthwash to your penis or your anus."[19] May I go out on a limb and add: don't put it in your vagina either.

While men who have sex with men are disproportionately affected by gonorrhea, the reasons for increased infections can be partly explained by a greater number of sex partners and higher rates of partner change. This would appear to support the popular wisdom that gonorrhea and other STIs are equal-opportunity infections driven by promiscuity. Unfortunately, this simply isn't true. In the United States, gonorrhea infects with racist, classist, and misogynistic tendencies (think certain politicians as STIs). The root cause of these disparities isn't found in the bedroom but instead in institutions where sex isn't even allowed: our nation's jails and prisons.

Explicit Bias

While most patients come to the STI clinic alone, it's not uncommon for a couple to walk in wanting to be checked out together. Sometimes both members of the couple want to be tested before deepening their sexual relationship; other times one or both is experiencing symptoms. Twosomes are most frequent, but I've also seen an awkward threesome: a patient with chlamydia who asked two of his partners to come in for treatment on the same day. Each partner learned of the other's existence when they met in the clinic's lobby (not a pretty scene).

When a couple presents together, each member is seen separately by different clinicians. Separating people is key, because sometimes one person believes they are monogamous, while the other person knows otherwise. After the sexual history and examination, each clinician steps out of the room and confers to compare the stories we've heard, agree on any tests to order, and align the treatment plan once we've established what's going on.

On one Thursday afternoon, there were two charts in our in-box for walk-in patients, belonging to an African American couple named Darryl and Aisha. I grabbed Darryl's chart and led him into an exam room, while Maya, one of our nurse practitioners, went into a room down the hall with Aisha.

Darryl had just been released from prison after serving for over a year. Once he got out, his first reunion was with Aisha, whom he'd been dating seriously before he'd been arrested. They celebrated his homecoming by having sex over the next several days; by three days after his release, he began having discharge and pain with urination. He asked Aisha whether she'd had any symptoms. She hadn't noticed anything, but she was willing to come to the clinic with him to be checked out.

As Darryl and I started to talk, he made it clear that he had not had sex with anyone other than Aisha. I gently probed about sex in prison, but he was adamant that "no one messed with me, and I didn't mess with anybody." I took a sample of the discharge from the tip of his penis and swabbed it onto a slide. I suspected that he had gonorrhea, so I also took a bacterial culture plate and wiped a swab onto the surface. Then I made my way to the lab to perform a Gram stain, which would help me confirm my diagnosis.

Within a few minutes in the laboratory, it was clear that Darryl had contracted gonorrhea. I went back to the exam room and told him what I had found. I reassured him that this could all be easily treated with two antibiotics, an injection of ceftriaxone and a few tablets of azithromycin. Of course, we would treat Aisha too. His eyes narrowed, and he looked askance at me. "How long have I had this again?"

I hedged a little. "Gonorrhea can incubate anywhere between a day to two weeks before it shows symptoms."

He shook his head. "But I just got out, there's no way I got this weeks ago."

"I know," I replied. I explained that most likely, he'd been exposed in the last two to five days. Darryl sat silently thinking it over. Then his face hardened—he had pieced it together. While he'd been in prison, Aisha had slept with someone else, contracted gonorrhea, and had given it to him.

Darryl rose and walked back into the lobby to await his treatment. I found the clinician who had seen Aisha, and we stepped into an empty room together. I told her that Darryl had gonorrhea, and she confirmed that Aisha had been with another partner while Darryl had been incarcerated. Aisha was not planning to tell him what had happened while he'd been away, but now his symptoms and diagnosis had forced her hand.

At that moment, we heard raised voices coming from the waiting room. Aisha had stepped out of the nursing area after her treatment, and Darryl had confronted her. They stood there arguing heatedly while the rest of the patients looked on. Darryl spun around and stormed off. Realizing he hadn't been treated, I rushed into the nursing area and grabbed a small paper bag containing two antibiotics. I headed outside after him and called his name.

As Darryl stopped and turned, I took the bag and pressed it into his hand. "I know you're angry right now, but take these when you calm down. It'll help you feel better."

He shook his head and stood silently for a moment. Finally, he muttered, "Thanks," turned around, and walked off. It wasn't exactly the homecoming he had been expecting.

The United States currently has the highest rates of incarceration in the world, dwarfing all other industrialized nations and our closest competitors: Rwanda and Russia. According to the Bureau of Justice Statistics, 2.2 million people are currently confined in our nation's jails and prisons, a population that is majority minority. According to the Sentencing Project, people of color make up 37 percent of the U.S. population yet make up 67 percent of the prison population. The racial disparities in incarceration are stark: one in three Black males born in 2001 will be incarcerated during their lifetime; Black men, like Darryl, are six times more likely to be incarcerated than white men.[20]

The devastating effects of incarceration on an individual's life are easily visible: loss of freedom and productivity, and the fracturing of relationships with partners, children, extended family, and social networks. Then there is the legally sanctioned discrimination that haunts felons after release, in the arenas of housing, education, employment, and voting rights. Because these collateral consequences overwhelmingly affect communities of color, they've earned the moniker "the new Jim Crow."

Removing men from communities has other unseen effects for heterosexual women, creating sex imbalances that reduce the number of eligible men available to have relationships.[21] This wouldn't matter so much if people could move easily between geographic locales to find desirable partners. No eligible men here? Let me hang out in the next town over to look for a mate. While this is theoretically possible, it isn't the reality, particularly in poorer communities with lack of access to transit options.

The way people choose sex partners is also not random. The law of attraction, or "like attracts like," strongly applies when it comes to sexual mixing patterns. We often choose partners who are like us in race/ethnicity, education, and socioeconomic status. In a study by Sevgi Aral of the CDC, she found that most Black and white participants reported sex partners of the same race/ethnic background in the past three months. This was most pronounced for Black women: over 90 percent reported only having partnerships with Black men.[22]

If Black women tend to seek partners within their race, yet there are not enough eligible men to go around, this results in a dating experience that Michele Andrasik from the University of Washington referred to as "sexual decision making in the absence of choice."[23] Emily Dauria at the University of California–San Francisco interviewed women in predominantly Black neighborhoods in Atlanta to see how this absence of choice affected their relationships. In neighborhood B (where less than 1 percent of males were incarcerated), single Black women lamented about a shortage of high-quality men, but the reason attributed to the shortage will sound familiar to all single women: all the good ones are married (or gay).

Meanwhile in neighborhood A, 12 percent of adult males were incarcerated, contributing to a sex ratio of seventy men for every one hundred women. Women in neighborhood A felt that the remaining available men were not desirable as long-term partners, and that multiple and concurrent sexual partnerships for men was a neighborhood norm. This led to the women having shorter partnerships that were focused on sexual activity, sometimes involving sex in exchange for financial or material support.[24,25]

These types of partnerships can profoundly affect sexual networks. As we learned from chapter 5, multiple concurrent relationships in a geographically contained space can contract a sexual network into a relatively closed structure, a situation ripe for the propagation of STIs. As a result, Black women nationally experience higher rates of infection with HIV and STIs—for example, rates of gonorrhea are almost eight times higher among Black women than their white counterparts.[26]

It would be democratic to think that gonorrhea is distributed randomly throughout the population. In reality, where and who you mix with plays a bigger role in whether you catch gonorrhea than the number of people that you sleep with. For this reason, network researcher John Potterat refers to gonorrhea as a "neighborhood" disease. In Colorado Springs, Potterat found that people with gonorrhea were mostly young and nonwhite, with four-fifths of cases concentrated in just one-fifth of the city's census tracts. Out of more than three hundred drinking establishments in the city, half of the people with gonorrhea socialized at the same six bars/clubs (a claim to fame not likely to be promoted on Yelp).[27]

Potterat analyzed sexual networks in the city, grouping sexually connected people into "lots," plotting the size of the lot, when individuals entered and exited, who became infected with gonorrhea, and where they lived and socialized. He found that the intersection of people's sexual and social connections within defined geographic spaces created a sexual "cloud," a perfect storm for transmission of gonorrhea in certain sexual networks in specific locales in Colorado Springs. He estimated that the risk of contracting gonorrhea by having sex within these clouds was three hundred times higher than in the general population.[28]

We can try to identify these clouds or hot spots of gonorrhea and increase access to STI testing and treatment in high-risk areas. But while the forces of mass incarceration continue to distort sex ratios in communities of color, we won't be able to eliminate the racial disparities in gonorrhea and other STIs. So while we advocate for health equity and criminal justice reform, we must bolster the public health safety net to provide more sexual health services, particularly in areas hard hit by STIs. It is just a matter of time before antibiotic-resistant gonorrhea strikes within the United States, and we can't afford to be caught with our pants down.

9

PrEPared

The Little Blue Pill Revolutionizing HIV Prevention

Here Come the Whores

In 1960, there were thousands of exalted naked women sitting on doctors' desks all over the United States. They were gold-painted plastic paperweights of Andromeda, Ethiopian princess of Greek mythology, shown breaking free of the chains that bound her wrists to a rock in the sea. These trinkets were gifts from Searle, manufacturer of the drug so widely known that it needed no name. Simply referred to as the Pill, marketing ads emphasized its ability to emancipate women from the irregularities of their menstrual cycle.[1]

By 1965, 6.5 million women had instead flocked to it to be released from their fears of unplanned pregnancy. The Pill would help fuel the sexual revolution of the 1960s, transforming cultural norms of sex into something for pleasure and not just for procreation.[2] But women's newfound sexual freedom was accompanied by societal fears about its power. By 1966, *U.S. News & World Report* reported widespread concern that the Pill would provide a "license for promiscuity." Could it unleash the hidden beasts within millions of women and ultimately lead to "sexual anarchy"?[3]

In 2012, another pill presented the nation with similar potential for promise and peril. This time it was Truvada, a nineteen-millimeter, sky-blue oblong tablet, imprinted with the number 701 on one side and GILEAD (the manufacturer) on the other. Each pill contained two drugs, emtricitabine and tenofovir, previously used for treatment of HIV. This time, Gilead Sciences asked the FDA to approve Truvada for HIV-negative people to take as prevention, in a practice that would become known as PrEP (pre-exposure prophylaxis). Just like the birth control pill, PrEP could be taken daily for as long as the person desired protection.[4]

The drugs in Truvada both inhibit reverse transcriptase, the enzyme HIV uses to replicate itself. The first of this class of drugs, AZT, was licensed back in 1987. While AZT is no longer used today, the reverse transcriptase inhibitors are still a mainstay of treatment more than thirty years into the epidemic.

Reverse transcriptase is a logical target to attack, since it's essential for HIV's survival. HIV can usurp our immune system to do its bidding, but it can't establish its presence without multiplying itself. It does this by copying its genetic code of ribonucleic acid (RNA) into our DNA, the basic building block of human genes.

A person's DNA is like a huge book divided into twenty-three chapters, or chromosomes, filled with three billion pairs of letters (A/T and C/G) called *nucleotides*. Once HIV enters a cell, its reverse transcriptase grabs these A, T, C, and G nucleotides from inside our cells and strings them together to make a single strand of DNA that matches its own genetic code.

The drugs emtricitabine and tenofovir in Truvada closely resemble our own A/C/T/G nucleotides; like underage drinkers with great fake IDs, it's hard to tell that they aren't the real thing. When the drugs are present, HIV's reverse transcriptase grabs them right along with our real nucleotides and adds them to the growing chains of HIV DNA within the cell. But if either emtricitabine or tenofovir is added to the chain, their structures are such that nothing more can be attached. The chain of DNA terminates, and so does HIV transcription.

It's reasonable to assume that drugs that prevent HIV from transcribing its genetic code would work if they were around *before* HIV actually set up shop inside our cells. Still, the thought of giving perfectly healthy people HIV medication, possibly for years, was rather radical. For the FDA to buy into it, the benefit for HIV protection would need to far outweigh any risks.

The evidence mounted in PrEP's favor. In 2010, the multi-national iPrEx study (Iniciativa Profilaxis Pre-Exposición) showed that among men who have sex with men and transgender women, Truvada reduced HIV infections by 42 percent.[5] Among people who actually took PrEP every day, its efficacy was 99 percent. Then in 2011, the Partners PrEP Study of African heterosexual couples (one partner HIV positive, the other negative) showed that Truvada decreased new HIV infections by 75 percent.[6] It was enough for the FDA to give the go-ahead to approve Truvada for PrEP in July 2012. For HIV prevention enthusiasts, it was a home run. Early adopters posted about their elation online: "Taking Truvada changed my life." "PrEP is a godsend."

A predictable backlash ensued. Journalist David Duran unwittingly gave it a name in his famous 2012 *Huffington Post* piece, "Truvada Whores," where he commented that the FDA's approval of PrEP was "encouraging the continuation of unsafe sex and most likely contributing to the spread of other sexually transmitted infections."[7] Duran would later have a change of heart, but the "Truvada Whore" had already gone viral (no pun intended), coming to represent the fracture within the gay community around PrEP.

On Duran's side of the PrEP divide were men who'd lived through the early AIDS epidemic, when condoms or abstinence were the only thing separating gay men from certain death. Larry Kramer, pugnacious longtime AIDS activist, once likened sex without condoms as akin to murder. Twenty years into the epidemic, at a speech at Cooper Union in New York City, he was the unswerving superego for safer sex:

> And by the way, when are you going to realize that for the rest of your lives, probably for the rest of life on earth, you are never going to be able to have sex with another person without a condom! Never! Every time you even so much as consider this I want you to hear my voice screaming like crazy in your ears. STOP! DON'T! NEVER! NO WAY, JOSE!

In addition to witnessing his friends and lovers die from AIDS, Kramer suffered frequent side effects from taking medications for his own HIV. The idea of a healthy person taking Truvada instead of using condoms was horrifying to him: "Anybody who voluntarily takes an antiviral every day has got to have rocks in their heads . . . There's something cowardly to me about taking Truvada instead of using a condom."

Michael Weinstein of AIDS Healthcare Foundation was also firmly in Larry Kramer's camp. He had started as an activist in the 1980s, providing hospice care for the dying. He was now the CEO of a global enterprise providing HIV care for more than six hundred thousand of the living. Along the way, his domineering style earned him several unenviable labels, including *Satan*, *thug*, and *enfant terrible*. He'd already generated ire two years earlier when he had his lawyers petition the FDA in an attempt to block Truvada's approval. Then in April 2014, he told the Associated Press, "If something comes along that's better than condoms, I'm all for it, but Truvada is not that . . . Let's be honest: It's a party drug."

Weinstein's protests aside, in May 2014, the CDC issued its first national recommendations for PrEP, stating it could benefit five hundred thousand Americans at risk for HIV.[8] A week after the announcement, journalist Donald McNeil inquired in *The New York Times*, ARE WE READY FOR HIV'S SEXUAL REVOLUTION? But the revolution had not yet arrived. At that time, the CDC estimated that a mere twenty-one thousand people, or less than 0.5 percent of those eligible for PrEP, were actually taking it.

The CDC upped the ante. In November 2015, they broadened their estimates to state that 1.2 million Americans, including one in four men who have sex with men, could benefit from PrEP.[9] Weinstein responded with his own provocative ads: "Open Letter to the CDC: what if you are wrong on PrEP?" and "PrEP: The Revolution That Didn't Happen," where he asked, "How long will it take for CDC to catch on to the failure of this strategy?"[10]

So which was it? The miracle that the world had been praying for or a corpse in the graveyard of failed HIV prevention strategies?

It depends on whom you ask.

Go Tell It on the Mountain

Clad in their briefs, Noah and his date entered the bathroom, each taking their place at one of the double sinks. Noah handed his companion a shiny square envelope. Keeping one for himself, he carefully tore it open. The HIV test cartridges held within resembled miniature white popsicles. Extending from the bottom of the popsicle stick, there was a tiny absorbent pad, round like the bottom of a tear drop. Holding the small plastic rectangle on the other end, each man swept the pad over their upper and lower gums, placing them into the vials of liquid resting on the counter. They locked eyes for a moment. Noah started the timer on his phone.

Twenty minutes.

Leaning with both hands against his bathroom sink, he watched time tick down. He focused his eyes on the small plastic cartridges and tapped his bare foot against the cool white tiles.

He'd been here before, but much had changed. Back in the mid-1990s when Noah was in his twenties, he was having a lot of sex but couldn't stomach getting HIV tests at his regular doctor. He'd had some luck with anonymous testing sites, but he always felt like he should be testing more frequently. In 1996, when the first home HIV tests became available, Noah went out and bought four, and always kept a stock of tests on hand. At fifty dollars a pop, it was a huge expense for a student on a budget, but he found it was the best way to manage his anxiety. If a test was positive, he wanted to be the first to know, to spare someone else from having to deliver him the bad news.

He'd prick his finger with a lancet and then blot it over and over to fill the circle on the test card with his blood. In those days, he'd have to send the test card off in the mail to be tested at a remote laboratory. During the two-week wait for his results, he'd have constant chatter going through his mind. *This is it,* he'd think. *It's going to be positive. It would be my own fault. I haven't always been safe.* He'd replay what it would be like to tell his mother that he was HIV positive and then envision himself in the future, standing in the same spot, his body ravaged by AIDS.

When his negative results arrived, his jaw would unclench,

his shoulders relax. He'd look up, thanking God and his parents for praying for him all these years. It was still working.

This was Noah's reality for over two decades. He'd test after having unprotected sex. He'd test every time he had flu-like symptoms, even if he'd been abstinent while studying for a big exam. Sometimes he'd test four times in a year, sometimes when he was particularly active, he'd test every month. He repeated this cycle of testing and worrying more than one hundred times during his twenties and his thirties.

Today was different. Since he had started taking PrEP, he knew that his HIV test would be negative. Like other patients taking PrEP, he tested for HIV and STIs at his doctor every three months—he didn't even need to test today. He was doing it for his date, so they could both be relaxed when they had sex. They had made a big deal about the testing, even planned for it. By force of habit, Noah still kept some HIV tests at home. They were still expensive, but he had a good income now. He offered his date a test from his personal supply. "Hey, you're worth a test," he ribbed.

It was time. He glanced down at his test cartridge. Seeing only one line in the test window, he smiled. Both of them were negative. He leaned in, placing his hand on his date's shoulder. Let the fun begin.

If you believe in statistics, then Noah should have been HIV positive years ago. As an African American gay man, his lifetime risk of being diagnosed with HIV is one in two, higher than that for Latino (one in four) or Caucasian men who have sex with men (one in eleven).[11] If his race and sexual orientation weren't enough, there was his ex-lover Terry of twelve years, who was HIV positive. Terry was usually good about taking his meds, but after so many years together, neither of them was vigilant about using condoms during sex. Then he and Terry would have an occasional three-way or four-way tryst when the opportunity arose. Those encounters usually involved drinking and little conversation, so HIV status and testing habits were not always discussed. Somehow Noah had managed to stay on this side of fate and was determined to remain there.

Blessed with good genes and disciplined about going to the gym, Noah is in his late forties but looks a decade younger. Although he's a consummate professional at the office, behind closed doors he's unabashedly open, including about his sex life. He lacks an inside voice, and his office walls are thin. At one point, his assistant pops in because he has said *anal sex* at high volume too many times. The chief financial officer of his company is headed over to talk to him (presumably *not* about his sex life), and his assistant pleads with him to keep it down.

He won't be deterred. "There's no one who I've let into my life at this point who can't know my details," he explains. "I've had a successful career *and* I have a social life. And I deserve to live."

Noah's of the opinion that God sent PrEP and that we ought to be able to enjoy the Lord's gifts. "When I first started PrEP and heard about groups in Los Angeles [Weinstein's AIDS Healthcare Foundation] coming out against it, I felt a need to acknowledge that I'm fortunate enough to be alive and be a part of the cure that God sent for this disease. For years, people have been begging, *Please, God, bring me something for HIV.* So it's not *the* cure, but these organizations can't stigmatize a twenty-one-year-old kid who is just coming out and dealing with enough already, simply because they don't think PrEP is the right solution. It's hard enough being a young man of color who is attracted to other men, we can't stigmatize them for doing something that actually works to prevent HIV."

In Noah's ideal world, he'd offer PrEP liberally: "If you are attracted to men and think that possibly, someday, maybe, you *ever* would do something with other men, you should be on PrEP."

Around his home in Minneapolis and while traveling on business trips, he had searched in vain for billboards touting the miracle. Why wasn't PrEP being advertised all over the place? he wondered. With his booming voice and his legacy as the son of a Pentecostal pastor, Noah made an ideal preacher for PrEP, and he began talking to everyone about it, even in casual conversation.

When a friend of a friend came to crash at his place, a gay twenty-four-year-old African American investment banker whom Noah didn't know well, he was determined to spread the good word. So Noah asked casually, "Hey, are you on PrEP?" His guest was taken aback. "Wow, you guys get really personal up here." He didn't say anything further. After being out of the closet and with an HIV-positive partner for so many years, Noah was such an open book that he had expected everyone else would be the same—until he realized why it might be awkward to talk about it.

"Here I am, just being so casual and thinking this is an opportunity to talk to a young man about PrEP. It didn't dawn on me that asking someone if they are on PrEP is essentially asking someone if they are HIV positive. Because if they are *not* taking PrEP, that might be because they are already HIV positive, and I would never just ask someone point-blank about their HIV status if they didn't want to talk about it."

He tried again at a dinner party in New York with seven other well-to-do gay men, a mix of friends and acquaintances. One guy who was HIV negative was talking about the latest research on PrEP. Several others at the table were against it; they felt the only reason someone would choose to be on PrEP would be to have unprotected sex. An argument arose, and in the midst of it, Noah mentioned, by the way, that he had started taking it.

"It was as if I had dropped a bomb. People just froze and looked at me, with glares that implied, 'Why would you say that out loud?'" One of his friends broke the silence: "You know, you are not supposed to talk about PrEP at dinner parties." Noah knew his friend was joking, trying to lighten the mood, but he was dumbfounded. *I mean, everyone in this room knows that I have sex, right?* he thought. "I wanted to keep going, but then someone finally said, 'Noah, stop. We are not going to talk about PrEP anymore.'"

He understood—it was complicated. "My generation was not 'supposed' to be getting HIV, because we knew better; we should have been using condoms." He also realized that someone might have been sitting there who was newly HIV positive. "There might be more guilt about becoming HIV positive [for that person], than for someone like my ex, who became positive decades ago. For the person who might have become positive in the last year, it's painfully clear: *I could have done something differently to prevent this.*" And he thought about how the older guys who had watched their friends die might mourn the fact that those friends could still be around if only PrEP had been discovered sooner. "My talking about PrEP could evoke feelings of loss, hurt, or guilt, so I realized that it's not okay to talk about it so casually anymore."

For a while, he kept his cards a little closer to his chest. Then Christmas came. Noah's family was hundreds of miles away in Pennsylvania, so his retired colleague Leann and her family had adopted him for the holidays, inviting him for Christmas dinner at their country club. Leann always fussed over him like a mother hen, rattling off a litany of questions: Was he sleeping enough, going to the doctor, eating well? Her measure of this was, "How many steaks have you had this week?" She always called him out if she thought he had lied about anything.

This was the year he had decided to pay back their hospitality and host everyone at his home. With the five of them gathered around the table, Leann began with her usual opening question.

"Noah, are you taking care of yourself? Are you going to the doctor?"

He froze. "Oh my God, am I going to tell this older couple, my adoptive parents, that I'm taking PrEP?" For a rare moment in his life, he was at a loss for words, and his silence caused immediate concern. Leann and her daughter both spoke at once, demanding to know what was going on.

By now he was laughing, and he finally said, "I must preface this by saying that we are about to have a personal conversation that I don't think you are prepared for." While they *were* really close, topics related to his sex life weren't usual fodder for their country club Christmas dinners. Leann's kids knew that his ex was HIV positive, but he hadn't even told Leann, because he knew she would just fret about him even more.

"By this point, what has come to my mind is that this is not a bad thing," he said. "I told myself, *You are going to share some information with friends of yours that you are close to because you know they have this fear for you, that you will become HIV positive. You are not going to give them a cursory answer. You are going to relieve them of this fear in a way that is going to invite them to celebrate your life with you.* So that's exactly what I did. And it was perfect."

While PrEP has been a gift to Noah and other men who have sex with men, in other parts of the world, those affected by the HIV epidemic haven't been so fortunate. In Africa, the burden of infections rests with girls and women, who account for almost two-thirds of the twenty-five million people living with HIV. For HIV researchers, PrEP's wins with the iPrEx and Partners PrEP trials instilled hope that it could also help the world's most vulnerable young women in sub-Saharan Africa. If PrEP was so efficacious for men who have sex with men and HIV-discordant couples, it should also work for African women.

Shouldn't it?

A Thousand Small Cuts

Jeanne Marrazzo smiled slightly as she hung up the phone with Mike Chirenje, her study cochair in Zimbabwe, half a world away from her office at the University of Washington. Their teams in Uganda, South Africa, and Zimbabwe had exerted Herculean efforts to enroll more than five thousand women into their study, and the National Institutes of Health (NIH) in the United States had committed over $100 million in funding. The Vaginal and Oral Interventions to Control the Epidemic (VOICE) trial, the world's largest clinical trial of HIV PrEP in women, was sailing along smoothly.

In VOICE, HIV-negative women were randomly selected to use one of three different prevention products: either a vaginal gel or one of two types of oral pills. One of the pills was Truvada, while the gel and the other pill contained one of its component drugs, tenofovir. In this case, women would use one of the products daily to prevent HIV, regardless of whether or not they had sex that day. As is customary in a clinical trial, about half of the women were unknowingly assigned to use placebo gel or pills. The study was a beast, with fifteen study sites and a thousand moving parts; Marrazzo and Chirenje had to communicate constantly to keep each other abreast of developments.

They had just ended a conference call with one of their study sites, and both were quite pleased. Their team reported that women were showing up to their study visits and using the products consistently. This was based on the participants' accounts and the amount of unused gel or pills that they would return at each visit. The women were also happy with the other services they received as part of the study: contraception, monthly counseling, and HIV testing, plus testing for other STIs. In areas of sub-

Saharan Africa where health care resources were scarce, these were huge incentives to get participants in the door. Women continued to vote with their feet—over 90 percent of them were showing up for study visits. Each participant could be enrolled for up to three years, and it was gratifying to have a strong showing out of the gate.

As is standard for a clinical trial, Marrazzo and Chirenje had scheduled regular meetings with a Data Safety and Monitoring Board (DSMB). The board was monitoring VOICE to make sure that there were no red flags that would end their study early, like unanticipated toxic side effects or other unusual outcomes. So far, none of the prior four meetings had raised huge concerns. Both drugs in the gel and pills had been used for years in people living with HIV and had an excellent safety record. In September of 2011, however, seven months after they had finished enrolling their last study patient, they were summoned to an early DSMB meeting at the headquarters of the NIH in Maryland.

Neither Marrazzo nor Chirenje knew what to expect. Two other studies of PrEP in Africa had recently been stopped, each for very different reasons. In April, the FEM-PrEP trial in women was stopped early for futility. In the initial review of the results, there were similar numbers of HIV infections in both the placebo and Truvada arms of the study, which meant that PrEP was unlikely to show any benefit even if the trial continued for its full planned duration. Three months later in July, the Partners PrEP study of HIV-discordant heterosexual couples was stopped because it was clear that PrEP was highly effective at preventing infection for the HIV-negative partner. For Partners PrEP, the board felt obligated to stop the placebo arm and offer PrEP to everyone because there was such clear evidence of its benefit. Maybe that would happen to VOICE too.

Marrazzo recalled the meeting with DSMB vividly: "It started with an unexpected, unpleasant, and unwelcome surprise." One of the two pills, oral tenofovir, did not seem to be effective at preventing HIV in their participants. Even if they continued following participants for the full three years, at the rate it was going, the board had determined that tenofovir would never demonstrate a significant reduction in HIV. The board recommended that they stop that arm of the study for futility.

It was a rude awakening, and Marrazzo and Chirenje were devastated by the news. Once the initial shock had passed, they had to figure out how to explain the failure to their study sites, whose staff had worked so diligently to enroll women and retain them in the study. They were doing everything right. Women were coming to their study visits and reported using the products. Why wasn't it working?

They tried to make their team feel better: "Maybe one of the oral arms didn't work, but the gel [and the other pill] would be okay." Two months later in November 2011, their DSMB convened another meeting. There were similar issues with the vaginal gel, and the board recommended stopping that arm of the study for futility. Then Marrazzo's anxiety set in: "Oh my God, this is really not good."

As she was recounting this chain of events, I recalled a scene from *Monty Python and the Holy Grail*, when the Black Knight has one arm chopped off by King Arthur. Blood is squirting from his torso, but he maintains his bravado: "'Tis but a scratch . . . I've had worse." As his second arm is amputated and he's exsanguinating, he's kicking King Arthur with his remaining limbs, exclaiming, "'Tis a flesh wound." Like the Black Knight, VOICE was also down two limbs and had just the Truvada arm left. But by that point, Marrazzo already felt like VOICE was dying from several large cuts. One of her colleagues feebly tried to console her: "Jeanne, like the surgeons say, all bleeding stops eventually."

There was still one missing piece of the puzzle. As part of the study protocol, the women provided blood samples, 160,000 in all, to be tested later in the course of the study. These tests would detect how much vaginal gel or pills each participant had used. The medication contained within the gel or pills would be metabolized, and traces of it would show up in their blood. The greater the usage, the higher the woman's blood levels would be. The women, Marrazzo, and Chirenje had no clue what the blood samples would reveal. In March 2012, after the two arms of VOICE had failed and shortly before the United States' premier HIV/AIDS meeting—the Conference on Retroviruses and Opportunistic Infections—they learned of the results.

More than two-thirds of the women assigned to the study drugs had little or none in their system. Women had enrolled in the study, over 90 percent had shown up faithfully for months with empty pill bottles and boxes of gel, but they hadn't actually used them.[12]

And with that, VOICE was done.

Marrazzo paced around the house in her robe and slippers, muttering to her wife, "I can't believe they didn't use the products; I can't believe they didn't use the products." I asked her if, once the blood tests had confirmed her and Chirenje's suspicions, they had wanted to hide in their beds with the covers pulled over their heads. She said that while they may have wanted to, they didn't. "You just can't do that. You have to keep communicating, you have to talk with the press, you have to talk with other people and other investigators who were like, *What the hell happened?* Somebody actually asked me, 'Jeanne, what did you do to VOICE?' It was really tough."

They knew what had happened, but they still hadn't figured out *why*.

If VOICE was the corpse, Ariane van der Straten from RTI International was the coroner dissecting the truth behind its demise. In two companion studies, she and her team spent hundreds of hours recording women who had been part of VOICE in focus groups and one-on-one interviews. As they methodically listened to each recording, they searched for clues as to why things had gone so wrong.

At first, the focus groups and interviews were not hugely revealing. No one would actually admit that they didn't use the medications. Sure, women would acknowledge missing a pill here or there, or occasionally forgetting to bring the vaginal gel with them when they were traveling—but nothing that explained why VOICE had failed.

Slowly, the team kept at it, and one painfully obvious issue emerged. For the women in VOICE assigned to take either antiretrovirals (ARVs) or placebo pills, the study medication bottles were labeled with the words *Division of AIDS*, the local government sponsor of the study protocol. Since the 1980s, South Africa has been an epicenter of the global AIDS crisis; today, nearly one in five South African adults is living with HIV. Consequently government-sponsored programs have succeeded in creating widespread national awareness around ARVs and an understanding that they are meant to prolong the lives of people with HIV/AIDS.

On the other hand, the idea of a healthy person taking ARVs daily to *prevent* HIV was a foreign concept. Even the participants themselves, despite being counseled about the purpose of the study, didn't fully comprehend the concept. Thoko, one of Van der Straten's focus group participants, explained, "What I know is that ARVs are for people who are sick, why would they [researchers] give them to us even though we are not sick? I would not understand that."[13]

The stigma caused by the medication labels was substantial.

When women's lovers, friends, or relatives discovered the pill bottles, naturally some accused them of being HIV positive and trying to hide their infection. Some women responded by leaving their pill bottles in plain sight, to prove they had nothing to hide. Others had to take suspicious loved ones with them to public HIV testing centers and submit to extra tests to prove they were HIV negative or bring them to the VOICE study clinics so that study staff could verify the truth.

To make matters worse, at some of these HIV testing centers and other government clinics, the staff told women that the scientists were secretly infecting them with HIV, not trying to protect them. For women who already had hesitations about using HIV treatment for prevention, the local medical establishment was reinforcing their fears of being treated like the researchers' guinea pigs.

As a participant by the name of Lilly said, "You know, it's scary to hear that you will take tablets meant for HIV-positive people, if you know very well that you don't have it. So that won't just be easy on you, even when you tell someone else that you are taking these kind of tablets, they won't understand, and they will think that you are lying and you have the disease or maybe at the clinic they will infect you with it because they are using us to test, and their question was why don't they test this on animals?"

These negative experiences weren't universal. Some women were lucky, and their lovers and friends encouraged their participation. But even some women with strong social support simply became bored of taking pills or using gel every day.

In any clinical trial that is supposed to continue for years, half of the battle is keeping participants engaged even at the moments that they tire of being part of the study. Paying participants or offering incentives can help keep momentum going.

What can also help is the participants' and staff's shared sense of community and purpose that comes from working on a study together over an extended period of time.

VOICE was hugely successful on that front. Very few women dropped out over the course of the study, so when a woman came to the clinic for a study visit, there were often plenty of other participants to talk to and share experiences with. Unfortunately, these waiting room discussions included women disclosing to each other that they weren't taking their pills or using their gel. In fact, nearly all 102 women Van der Straten's team spoke with had overheard another participant admitting that they hadn't used their study medication. In the same way that peers can help motivate you to stick to a routine, they can also just as easily do the opposite—"Let's scrap the exercise and grab a doughnut; everyone else is doing it."

While women had pointed the finger of blame at each other, they still hadn't really owned up to their own actions. Recall that all the women in VOICE were giving blood specimens throughout the study. Van der Straten and her team decided to show the women their blood test results. To do this, they represented the test results with photos of teapots next to five teacups. Five full cups of tea would signify that their average blood levels of medication were high; the number of full cups would decline based on lower blood levels or less frequent product use. It was a different approach to a usual study focus group; a woman could neither deny nor augment her medication use, since the results were staring her right in the face. "Here are five empty cups—there was no study drug in your system. Can you tell me what happened?"[14]

With this approach, most women with low adherence eventually admitted that they didn't use the products. Some assumed that there would be enough of the others who were using them that their individual actions wouldn't make a huge impact. Ul-

timately, many hoped and believed that despite their lack of adherence, the products would be protective against HIV. So they
dumped out their pills and gels, told the VOICE staff what they
wanted to hear, crossed their fingers, and hoped for the best.

Women like Nyaradzo were horrified when they learned
VOICE was being stopped for futility: "I felt sick because I was
telling myself that it does work and even now I wish that . . . they
can get another thing to prevent HIV. . . . But I still feel that . . .
it is just that we were not using it properly. That's why they found
that it is not effective. . . . Because the thing is I also know that
sometimes I didn't insert it [vaginal gel]."

From Jeanne Marrazzo and Mike Chirenje's ten-thousand-
foot view as the study chairs, VOICE had died by several large
blows. First, one and then a second arm of the study being
stopped for futility, plus the blood test results that put the final
nail in the coffin. Ariane van der Straten's postmortem is valuable because she was able to tease out the women's feelings in the
context of other social influences. She had the bird's-eye view of
what Marrazzo and Chirenje could not see—that VOICE's demise was actually a death from a thousand small cuts.

Sure, Marrazzo was disappointed about the years of time
spent to secure research funding, countless hours to get the trial
off the ground, the $100 million spent. But she also felt betrayed.
It was like being in a relationship with a woman for months, then
discovering that she was only pretending to like you because she
didn't want to hurt your feelings—then multiply that by several
thousand women.

It had become clear that women wouldn't use those particular
prevention products, but surely they didn't want to have HIV. So
now what? Before VOICE, Marrazzo had spent a decade of her
career studying vaginal microbicides, antiviral lubricants to be
used during sex that could kill HIV on contact. Most of those

lubricants hadn't panned out the way she'd hoped either, but each failure was giving her greater insight.

"I really think you need to meet women where they are. The women we were working with in the highest [HIV] incidence settings were simply not in a place where they were thinking, 'I'm going to have sex with this person, and I could get HIV tonight.' They were in a place where they were thinking, 'I could get pregnant tonight.' HIV was just not part of their day-to-day."

Certainly, women in Africa are knowledgeable about the threat of STIs and HIV. But many also struggle with lack of employment, hunger, violence, unstable housing—issues that intensify immediately with an unplanned pregnancy. Something like HIV, an infection that lies silent for decades, could easily fall lower on their list of concerns.

The failures of VOICE and similar trials like FEM-PrEP demonstrated that HIV prevention with daily pills or gels wouldn't fly with many African women. PrEP has gained much more traction among gay men, but there's one crosscutting issue in both populations. For PrEP to work, one must get on board with it before they become exposed to HIV. Unfortunately for some, that realization comes after they've already missed the boat.

The Blind Side

My friend Ronan was born and raised in South Africa, but after medical school, he'd lived all over Europe before finally deciding to settle in New York. His accent is difficult to place; his patients can tell that he's from elsewhere, but they generally give up trying to pin it down beyond his last country of residence.

One day, he was in town for a conference and we decided to meet at a local deli beloved by the carnivore hipster set—its signs boasting, YOU CAN'T BEAT OUR MEAT. Ronan arrived dressed to the nines, his tailored suit a stark contrast to the hoodies and jeans of the mostly male lunch crowd. His dapper dress was the cherry on top of a combination of his adorable and devilishly funny personality. He was living proof of the local hetero women's common lament—all the good ones are gay.

With pastrami piled high on my plate and a sensible salad on his, we settled in and started chatting. He was smiling, but he wasn't his usual self. He asked me to think back to the last time we had met in person. We had both been at a scientific meeting in Seattle. The meeting had ended, and I had run into him at an outdoor café near our hotel having a drink, sunning himself in shorts and a T-shirt. We chatted about his boyfriend, Brian, back home in New York; things were getting serious and they were talking about getting married. I remember feeling happy for him, exchanging pleasantries, and basking in the warmth of the afternoon.

I didn't know it at the time, but Ronan and Brian had been giving each other some latitude when it came to their sexual relationship. While they were home, they slept with each other exclusively. But when either of them traveled for work a few times a year, they were free to be with whomever they chose. When they returned home, before having unprotected sex with each other, they would be tested for STDs and HIV, then resume having sex once they had received negative results. This was an inviolable pact between them and had become so routine that they didn't even have to discuss it anymore. In more than two years, neither of them had ever contracted an STI despite having multiple partners outside their relationship.

That evening after our visit in Seattle, while I had enjoyed delicious solitude in the hotel bathtub, Ronan had had a few drinks and then headed to a local sauna frequented by gay men. After slipping into his towel, he walked around for a few minutes before he locked eyes with a cute guy with sandy-brown hair. It didn't take much talk to establish a mutual attraction and interest in having sex. They quickly slipped into a shower stall and started removing each other's towels.

The guy gave Ronan oral sex for a bit but was quickly interested in going further. Ronan was versatile for anal sex, meaning he could either be the insertive ("top") or the receptive partner ("bottom"), depending on the guy and the circumstances. When having sex on the road, he generally preferred being the top, since he'd be less likely to catch HIV that way. And he would typically use a condom, as he did in this instance.

"So I was fucking him, and at some point, I looked down and saw that the condom had broken." Still breathing hard from effort, he touched the guy's shoulder to get his attention.

"Hey, the condom broke. What's your HIV status?"

"I'm positive," the guy said, "but I'm undetectable."

Ronan was relieved. If this guy was taking HIV meds and his HIV viral load was undetectable, then Ronan's risk of contracting HIV was essentially zero. Plus, he had been the top during sex, so his risk was even lower still. And the guy was hot, and they were having a great time; there was no turning back now.

When he returned to his hotel room late that night, he had a sudden thought and began rifling through his toiletry bag. There, alongside his condoms, he had also packed a starter pack of HIV medication similar to what he might prescribe to a patient *after* being exposed to HIV, a practice known as postexposure prophylaxis (PEP). It was another insurance policy in case a condom ever broke or fell off while away from home.

It was overkill, but what the heck, he thought, and popped

the pills just in case. He would continue taking PEP until he got his negative HIV test results.

He flew home from Seattle and went to work several days later. When he walked in, he grabbed his nursing manager and told her he needed a check-up for STDs and HIV. She promptly drew his blood, collected his urine, and took a long Q-tip swab and twirled it across the back of his throat and tonsils. This was the standard procedure to test for HIV, syphilis, gonorrhea, and chlamydia at all the places where he'd been sexually exposed.

Because his clinic was part of the health department, and he personally knew all the surveillance staff, he didn't want his actual name linked with any positive test results that were entered into the public health registry. So he always tested under the same pseudonym—let's call him Richard Smith. He used his own date of birth so he could be sure not to confuse the results. He listed himself as the ordering doctor.

He always checked on his own results, but he knew that it would take several days, so it quickly left his mind while he started to catch up on the backlog of work that had accumulated during his time in Seattle. A few days later, he was in a meeting with one of the HIV program staff when the phone rang. It was the laboratory calling with patients' test results. Usually, these calls went to his junior physicians first, but they were all occupied, so the call was transferred to his desk.

"Ronan?" It was David, one of the resident physicians working in the lab on his immunology rotation.

"Speaking."

"I've got a positive HIV result on one of your patients."

"Okay, who is it?"

"Richard Smith."

For a moment, there was still a chance that it was a different Richard Smith. As the resident read back the date of birth, which matched Ronan's own, the tiny shred of hope that he'd held quickly vanished. Later, he could appreciate the irony of being told his own positive HIV test results with his fake name over the phone. But at that moment, he felt all color drain from his face. He managed to hold on to the receiver. His voice was calm as he responded, "Okay, thanks for the details. Thanks for letting me know." He told the resident that he would order a confirmatory test—a Western blot—which would come through in two days. He knew deep down that it would also be positive.

As he hung up the phone, he looked up at his colleague sitting across his desk. He had known her for more than a decade, and he loved and respected her. He suddenly felt the need to tell her, to anchor himself to someone else as waves of dread began to wash over him.

Should I do it? he thought. *Can she tell something's wrong? Can she see the look on my face?*

"I'm sorry, Naomi," he said. "I've just been informed with a positive HIV result. And it's mine."

Her eyes widened, and a look of slight panic began to cross her face. She began fluttering her hands and speaking rapidly, clearly flustered by the news. "Oh! But I . . . I'm not a clinician. I'm not trained on how to do post-test counseling!"

He found it endearing that she wanted to say the right thing, even though there wasn't any right thing to say at that moment.

"Naomi, it doesn't matter. Fuck! I don't need post-test counseling." Another wave of realization hit him. "Fucking hell!"

"I don't know what to say."

"It's okay. There's nothing *to* say. This is what's happened."

They sat in silence for a minute as Ronan's head began to spin. How could he possibly have contracted HIV?

It had been just over a week since he'd had sex in Seattle, yet somehow his HIV test was already positive. He knew this recent test wasn't detecting a new infection from the Seattle trip. He had just tested out of habit as part of his post-travel routine. While the latest generation of HIV tests can detect a protein (p24 antigen) from the virus about two weeks after being infected, Ronan's lab was still using the classic HIV antibody test, one that detected antibodies from one to six months after being infected.

His heart sank as he thought about his test results, the sauna, the broken condom, the HIV medications he had tried to take to prevent infection. None of it had mattered. *I was HIV positive before that trip,* he realized. He racked his brain, but he was too overwhelmed to think straight. If he was positive before he went to Seattle, when had it happened? And how many times had he exposed Brian? *Shit.*

For just a moment, Ronan considered texting him, and then thought better of it.

Was he really going to tell him over the phone? There was nothing he could do about it now. He and Brian had slept together multiple times before Seattle, so Brian was already far beyond the seventy-two-hour window to take postexposure prophylaxis himself. Ronan would wait until that night.

When Brian came home from work that evening, Ronan shared his news. Then he called his best friend, who came over, and told him as well. The three of them spent the evening together in a state of shock. His friend and Brian were upset. Ronan sat there, numb, the weight of it yet to sink in.

The next day, he went to an HIV specialist he knew to check his T cell count and viral load. He'd worked in several of the hospitals in the area and worried he'd be recognized by the nurses and other staff. He gritted his teeth and went in, hoping to go unnoticed. He was quickly ushered into a private room and examined. The doctor ordered his medications and his blood work under the name Shirley Temple—Shirley's date of birth was one day later than Ronan's—so that he could track the results down later.

He had no detectable virus in his system, as he'd been taking prophylaxis medication since his tryst in Seattle. His T cell count was over a thousand—his immune system was fine. It turned out his HIV could be treated with one pill that combined three drugs, taken once a day. Twenty years earlier, it would have been closer to twelve pills a day. He felt a small bit of gratitude for this. He took his medication faithfully. He felt zero side effects. To his body, it was as if nothing had happened.

Meanwhile, Brian was tested and then tested again. He was negative.

He and Brian got married several months after Ronan's diagnosis. They were still trying to figure out how to maneuver their sex life with one of them being HIV positive, but so far they were making it work, and Ronan continued to do well physically.

His mental state was another issue.

"My drinking spiraled out of control. I was getting hammered every night. I was so unmotivated, not wanting to do stuff around the house, not wanting to socialize. I started seeing a psychologist, who told me I really needed to cut my drinking down for many reasons, but primarily because I was not going to be able to cope with this diagnosis if I was intoxicated all the time."

Every weekday morning for months after his diagnosis, Ronan woke up hungover and dragged himself into work to see his own patients. Many were gay men like him. Some professed to Ronan that they were acting like "sluts," sleeping with dozens of partners in a few months' time. Now that they were taking HIV PrEP, they were all enjoying sex with less worry of catching HIV. Every new guy he'd prescribed PrEP to was a reminder of his own bad luck.

He began to seethe.

"I admitted that to my shrink. I felt bad that I felt angry. I asked myself, 'How come you guys get to go and bang whoever you want, however you want, because you are getting PrEP?' I ended up with HIV, and it wasn't like I was doing a lot of risky things." In his entire sex life spanning almost twenty-five years, he'd only had one measly STI (chlamydia). He'd always been so careful, so he'd never thought he'd need PrEP. It didn't seem fair.

As a physician versed in the statistics of HIV transmission, Ronan knew he had defied the odds, but not in a good way. His risk of contracting HIV as a top for anal sex with an HIV positive partner was only 0.10 percent (11 in 10,000 exposures). That was much better odds than being the bottom (receptive partner) where the odds could be more than 1 percent per sex act (138 in 10,000 exposures).[15]

He had even tried to mitigate some risk through taking prophylaxis medications when his condom had broken. But he had obviously miscalculated his risk at some other point *before* then, contracting HIV in another situation that hadn't even registered to him as particularly high risk.

He eventually figured it out. In a cruel twist of irony, he'd been infected with HIV at the International AIDS Conference in Melbourne, a few months before his trip to Seattle. He'd met someone who'd told him that he was HIV negative, and Ronan had been the top for anal sex—part of the time without a condom. They'd parted ways hours later, none the wiser about what had transpired.

Had the guy lied to Ronan about his HIV status? Probably not. More likely, he just didn't realize that he was HIV positive. Negative test results can be falsely reassuring. It's not because HIV tests are inaccurate, it's that testing can be negative early on, and the results are only reliable until the moment someone goes out and has unprotected sex again. At that moment, the person's HIV status becomes *unknown* until the moment they test negative again.

It sounds great when someone says that they test every six months. But if they celebrate being HIV negative by having unprotected sex the week after getting their results, there is a long window until the next test (five months and three weeks—or 171 days—to be exact) where that person's HIV status is unknown. Instead of that person being "negative," their true status is more like "maybe." People may have the best intentions and *think* they are HIV negative, but the HIV epidemic goes on because every day there are people who are "maybe positive" having unprotected sex with one another.

What's a cautious person to do? Testing more frequently shrinks the window of time in between unprotected sex acts where HIV status is unknown. And new, more sensitive HIV tests can pick up infection within two weeks, also shrinking the window of time between becoming infected and getting a positive HIV test result. With more frequent testing and more sensitive tests, the window of doubt gets progressively smaller and smaller. But with testing alone, that window will always be large enough for HIV to make its way inside.

As a physician, Ronan knew that the best way to cover that window was to put up a screen. Condoms worked if people would use them, but for those who wouldn't or couldn't, PrEP was the strongest barrier to shutting out HIV. Despite his resentment and grief about his own situation, Ronan kept prescribing PrEP to help his patients, though he could no longer help himself.

Meanwhile, Ronan had struggled to keep his diagnosis a secret, and he hadn't seen a doctor for six months. His first pseudonym, Richard Smith, had worked well before the diagnosis, since Richard had always had negative tests and never actually had to interact with the health care system. Under his second pseudonym, Shirley Temple, he had encountered numerous difficulties picking up his medication because he didn't have any identification that matched Shirley's date of birth. He couldn't bear using his real name because he knew so many people at neighboring labs, hospitals, and pharmacies—and he was still fiercely private about his diagnosis. With the help of a nurse at a neighboring clinic, he managed to create a third identity for himself that allowed him to do everything he needed. He was finally back in HIV care, drinking less, and on the therapist's couch every week.

Our lunches half-eaten and abandoned, there was nothing left to say. I hugged him hard, which was enough for the time. Pulling back to say goodbye, Ronan grinned, a genuine smile with little crinkles on the corners of his eyes. He had another pressing issue. He and his friends were dressing up as the Golden Girls for Halloween. Could I help him find glasses that would transform him into the feisty Sophia? Reveling at the power of costume play to lighten the most serious of moments, I rattled off a list of local thrift stores for his hunt, hoping he would find what he was looking for.

Getting to Zero

If the world became perfect tomorrow, there would be no need for PrEP. In this perfect world, people who are already HIV positive would know their status and take antiretroviral medication, suppressing the virus to an undetectable level. HIV treatment is an effective strategy for prevention, because someone with an undetectable viral load is unable to transmit HIV.[16,17] This is the concept behind HIV treatment as prevention (referred to as TasP), and it could prevent HIV at both an individual- and population-level. Imagine if everyone with HIV had an undetectable viral load—the HIV viral load in the community would be close to zero, and there could be no new HIV infections.

The concept of undetectable=untransmittable (U=U) is more than just a tagline of the nonprofit Prevention Access Campaign. U=U is supported by evidence from three large studies including over one hundred thousand acts of condomless sex between serodiscordant couples—where one person is HIV positive and the other is negative.[18-20] Anthony Fauci, director of the National Institute of Allergy and Infectious Diseases at the NIH, commented on the studies' findings: while scientists don't like to use the word *never,* the risk of transmission from someone with an "undetectable viral load is so low as to be unmeasurable, and that's equivalent to saying they are uninfectious."

Not to dampen the mood, but unfortunately, one in seven Americans with HIV still doesn't know their status, and that's unlikely to change tomorrow. So TasP can make huge impacts, but it isn't sufficient on its own. What about condoms? In the perfect world, condoms would never fail, and everyone at risk would use them. But we in the real world know that condoms can and do fail. And we mortals also make mistakes; we cheat or get cheated on, get drunk and make bad choices, or get carried away by lust.

With the HIV epidemic approaching its fortieth anniversary, condoms, HIV testing, and TasP are all great tools—surely, we'd be worse off without them. Still, Ronan and the forty thousand Americans diagnosed with HIV every year are proof that there is too much vulnerability in our defense against HIV. We need something like PrEP.

Still, it's not the panacea. PrEP only works if people use it, and women in Africa still desperately need a product that they will actually use. Until 2019, there was only one option for PrEP

(Truvada), and it required daily dosing. Now there are two options for PrEP (Truvada and Descovy), as well as the ability to
use PrEP on demand (taking two pills before sex and two pills
after sex). And the menu is still expanding. Like its contraceptive predecessor, the Pill for HIV prevention is paving the way
for a wider menu of options—injectables, implants placed under
the skin, vaginal rings. Jeanne Marrazzo hopes that one of these
will work if the prevention community takes a different approach
with women: "Here's a package of services that makes your overall lives and health better. PrEP is just a part of it."

Then there's the hangover of the Truvada Whore. In September 2017, John, a thirty-five-year-old PrEP user from the
UK, updated his profile on the dating app Grindr to disclose
that he was on PrEP. "I can safely say I have never seen so many
guys suddenly lose interest or just stop talking/block me." On
a PrEP Facebook group, he asked, "Has anyone else dealt with
this? Made it clear with one guy that I take it as extra precaution, still use condoms, and he said, 'Didn't think you were "that
kinda guy"' . . . what?!" Stigma once reserved for HIV-positive
folks has reared its ugly head toward those trying to stay negative. But only if we let it. I messaged John to thank him for being
out about his PrEP use. The more people like him in the world,
the less stigma there will be.

Stigma notwithstanding, the PrEP movement has gathered momentum. The CDC reported that PrEP use increased
500 percent between 2014 and 2017, and that a third of men
who have sex with men in urban areas were using it.[21] In 2018,
CDC director Robert Redfield estimated that about two hundred thousand people were taking PrEP; it may have reached its
tipping point toward widespread acceptance. It's also affecting
more than just HIV prevention. Kim Koester, the anthropologist
from the seminal iPrEx study, touts the unexpected benefits of

PrEP she's witnessed: "I was so impressed with the tremendous impact, the psychological impact that was happening for the people that were using PrEP. I think PrEP definitely empowers people. That you can think about your future now without it being clouded by HIV, you might make [other] decisions that are healthier for you over the longer term."

Jonathan Volk agrees. He runs one of the largest PrEP clinics in the country at Kaiser Permanente in San Francisco. "I think it's probably improved a lot of relationships [between HIV-discordant partners], relieved anxiety, improved sexual satisfaction. I think a lot of folks have lived for many years with a lot of anxiety around sex."

Prior to PrEP, people from the HIV-positive and HIV-negative community were sometimes hesitant to have sex with each other. Naturally, this caused divisions in sexual and social networks based on HIV status. Koester and Volk both feel that PrEP has created bridges between the two. Volk mused, "I think that it opens up the community. The community was very much split for a long, long time and I think PrEP is helping fix that. Also the fact that we have treatment as prevention . . . I think the combination of the two [suppressing the virus among HIV-positive people and PrEP for HIV-negative people] is really changing the game."

While PrEP is helping us get closer to the perfect-world scenario for HIV, it's important to recognize that it may also be acting as lighter fluid for an STI epidemic that is already aflame. Along with our suspenders and axes, we'll certainly need our rubbers. There's still a big fire to fight.

10

When the Rubber Meets the Road

*The Lowly Condom and the
Siren's Song of STI Prophylaxis*

Building a Better Mousetrap

The University of California–Berkeley is famous for its love of protest: free speech and the Vietnam War in the 1960s, apartheid in the '80s, Occupy in 2011. These demonstrations often convene in a plaza at the foot of Sproul Hall, a large administration building with an imposing beaux arts façade. And it was sometime between the anti-Vietnam and apartheid protests that Sproul Hall served as the epicenter of a less political, albeit monumental movement: the first National Condom Week, launched by UC–Berkeley peer educators in February of 1978. Today, festivities in honor of the prophylactic are held by universities and nonprofit organizations all over the country.

On Valentine's Day 1994, I found myself addressing the lunchtime crowd at Sproul Hall to welcome them to the university's Condom Week activities, with giveaways of condoms, dental dams, lubricant, and information pamphlets. Ordinarily, I would have been a nervous wreck, but I had donned special armor for the occasion. I delivered the speech dressed as a giant condom (complete with reservoir tip), which left only my face exposed to the crowd.

While standing at the top of the stairs, I held a banana over my head and demonstrated proper condom technique: carefully pinching air out of the tip and rolling the condom slowly down along the length of the fruit. Then I demonstrated how it should be removed after ejaculation, holding the base of the condom and slipping it off so that the hypothetical semen would stay inside. Most people laughed and clapped, while a few in the crowd gave me dubious looks. It didn't faze me in the least. If a laugh at my expense might help someone get over their own embarrassment, it was well worth it to me.

My condom cosplay made such an impression that I wrote about it as the opening paragraph of my medical school admissions essay. One of my advisers tried to talk me out of it; he was a professor in his midfifties, and to paraphrase his advice, "Most admissions committee members are just like me. Trust me, they aren't ready for this."

Plenty of schools still invited me to interview, but I do think my statement may have given some committee members a bit of pause. During an interview at a prominent midwestern university, the interviewer perused my application and then asked me point-blank, "Have you ever used illegal drugs?" Not sure if that was a standard question or only asked of people who would admit to dressing up as a condom.

Scholars debate exactly how long the condom has been around, but Italian anatomist Gabriele Falloppio (discoverer of the Fallopian tube) published the first written accounts of its use in 1564. Falloppio's early condom was made of linen soaked in a chemical solution, designed to fit over the head of the penis and tied in place on the shaft by ribbons. After testing it with 1,100 men, he claimed that no one using his device had contracted syphilis, which was wreaking havoc across Europe at the time.

We've come a long way in sourcing materials since then, from animal intestines during the Renaissance to vulcanized rubber in the 1800s to the discovery of latex in the 1920s. In modern times, some desperate creative types have even attempted do-it-yourself (DIY) condoms: plastic sandwich bags, popsicle wrappers, plastic wrap, balloons. While I applaud folks on their commitment to prevention, I'm afraid nothing makes a good DIY condom substitute, particularly anything with sharp edges like a popsicle wrapper.

Although most of us will use condoms at some point during our sex lives, many of us have a fraught relationship with them. In one camp, there are condom enthusiasts like me who tout them as an effective prevention tool against STIs and HIV. Condoms save lives—of course we should love them. In the other camp, there are those of us who can't stand them for a number of reasons: they don't fit, they dull sensation, they're awkward and a hassle.[1]

Unfortunately, the haters camp is full of people at risk of STIs and unplanned pregnancy. In the United States, only one in three men and one in four women aged fifteen to forty-four used a condom the last time they had vaginal sex, and less than 9 percent of women used condoms as their primary contraceptive method.[2] Although teens are the age group most likely to use them, they're losing popularity with the generation Z set as well. According to the CDC's Youth Risk Behavior Survey, condom use by high school students during their most recent sexual encounter declined from 62 percent to 54 percent in 2017.[3]

Our dislike of the condom for contraception is not uniquely American. According to the United Nations, condom use among married women in the United States is on par with that of France,

Germany, and Canada, ranging between 5 and 15 percent.[4] But the condom does have its fans elsewhere. It is unquestionably big in Japan, where 40 percent of married women used condoms as their primary method of birth control, by far the highest rate among industrialized countries.[5]

What is so special about Japanese condoms? While they are constructed of the same source materials, the Japanese are known for producing the thinnest condoms in the world. While standard condoms are about 0.06–0.08 millimeters thick (about the width of a human hair), the Japanese make ultrathin versions that measure from just under 0.02 to 0.03 millimeters in thickness.

But there is a more practical reason for the condom's popularity in Japan: a lack of other alternatives. Japan was the last industrialized nation in the world to legalize the Pill in 1999, after it had been used for nearly four decades in other countries. According to Debbi Gardiner in *Wired*, women's groups and Japan's four female health ministers railed against the Ministry of Health and Welfare after they approved Viagra in six months, while applications from birth control pill manufacturers had been languishing for years.[6] But even after the government's approval, negative press about the Pill's side effects have kept Japanese women away. According to a 2016 study of national contraception practices, only 3 percent of Japanese women aged sixteen to forty-nine used birth control pills, compared to more than 80 percent who reported using condoms.

Americans may not love condoms as much as the Japanese, but they still buy them in droves. In the United States, the domestic condom market was valued at over $1.3 billion in 2016 and estimated to hit nearly $1.7 billion by 2022.[7] And condom makers are still tweaking their products to encourage more consumer use. Since the advent of latex condoms in the 1920s, condoms have become thinner, stronger, and more varied than ever, bestowed with textures, flavors, and colors galore. There are synthetic options for the latex-allergic made of polyurethane and polyethylene. And for the sexually active animal rights activist, there are even vegan varieties, lacking the casein typically used in condom manufacturing.

But until recently, there was little that companies could do to address one of the most common condom complaints: poor fit. While today there are brands that are more snug or roomy, for many years condoms came in one length and width, a by-product of FDA regulations. Meanwhile, penises come in all shapes and sizes. While the average penis length ranges from 4.7–6.3 inches when erect, there are outliers in both directions. Coming from someone who has seen a lot of penises (in a professional capacity), I can attest that their shapes are also as varied as the bodies that they are attached to. Some are long and lean, others short and round. If shoes come in eleven sizes and nine widths, why shouldn't condoms be the same?

Today, the FDA allows more leeway in condom sizing, so naturally someone decided to capitalize on a potentially untapped market. Today the Global Protection Corporation offers myONE Perfect Fit: custom-fit condoms in up to sixty different sizes. Customers measure their erect penis to determine its length and girth against a FitKit, a printable ruler with seemingly random letters and numbers representing various measurements. The company then sends along the designated number and letter combo, along with a few other sizes for good measure.

Other than fearing a papercut from the FitKit, writer Scott Muska gave the custom condom a thumbs-up in his piece for *Cosmopolitan:* "The fit was snug but not constricting. It was freeing—like latex was no longer attempting to kill my boner." But all this penile freedom comes at a price; custom condoms cost about two to three times as much as their standard counterparts. And there was still an issue with the condom that even the perfect fit could not solve. As Muska put it, "It was still a condom." Then he added, "Which, in my experience so far, means that there will always be less sensation than if I wasn't wearing one."

But what if that weren't the case? Imagine if there were a condom that actually felt better than bare skin against skin? In 2013, the Bill & Melinda Gates Foundation decided they'd cast a net and see what type of futuristic condoms they might scoop up. They offered seed grants of $100,000 to innovators who might develop a "condom that significantly preserves or enhances pleasure, in order to improve uptake and regular use."[8] The strongest candidates would be eligible for $1 million more to assist in development and scale-up. Out of 953 applicants who threw their hats into the ring, 11 candidates were chosen, employing materials such as graphene and nanoparticles, or novel applicators to eliminate the hassle of correct placement.

Even with the Gates Foundation's funding, they knew that it could take years before a new condom technology might end up in people's hands. While they were making their call for new condom candidates, the foundation also threw their support behind other products that were already in development. The most promising of these was the Origami Condom, a clear sheath made of silicone and shaped with accordion folds, which came in three models: one that could be pulled on the penis like a mitten, and others that could be worn internally in the vagina or anus. The accordion folds afforded more sensation and easier application onto the penis than a rolled condom. In the foundation's blog, *Impatient Optimists*, they touted this condom's great potential: "Origami Condoms provides an excellent example of a private enterprise focused on new condom design," a design dedicated to "emphasizing the sexual experience."[9]

The Origami Condom's developer, Daniel Resnic, seemed like a perfect person to disrupt the condom status quo. He had contracted HIV from a broken condom in the 1990s, which launched his personal crusade to create a better prevention device. As Resnic's star began to rise, his condoms were featured by multiple major media outlets, including Fox News, *The New York Times, The Atlantic,* NPR, the BBC. He gave a TEDx talk on how the condom industry had duped us into accepting a substandard prevention device and how his product was poised to end that once and for all.

The Origami Condom was so promising it even gained the federal government's attention; the NIH awarded him $2.4 million to develop and pilot test his product. And Resnic was planning to seek FDA approval specifically for anal sex, an indication lacking for standard condoms. In multiple media outlets, Resnic estimated that the Origami Condom might be on the shelves as soon as 2015.

Unfortunately, by May of 2014, Resnic's metaphorical condom had broken once again. First reported by Elizabeth Harrington of *The Washington Free Beacon,* Resnic was accused by one of his former employees of spending the NIH's funds (a.k.a. our tax dollars) on activities that appeared a far cry from legitimate research: personal vacations, tickets to a party at the Playboy Mansion, full-body plastic surgery, and a condominium in Massachusetts.[10]

Resnic turned around and accused the employee of the wrongdoing and proceeded to slap a lawsuit against him. Resnic suggested that the employee pay back the NIH's money, proposing a bizarre payment plan: "Make restitution to my company of the stolen monies ($487,377.32) at one dollar ($1.00)/week, by personal check, sent by U.S. mail, until the funds are recovered. (Yes, that's 487,377 weeks or 9,372 yrs. of payments)."[11]

A judge dismissed Resnic's lawsuit against the employee. Resnic was then on the hook for the misappropriated money, and he realized that he'd need to negotiate settlement terms with the NIH. It seems that Resnic will be tangling with the government for some time to come. Meanwhile, the Origami Condom's website and Indiegogo pages are down, and without further investment, it appears that Origami Condom won't be appearing on shelves anytime soon.

Unfortunately, it's uncertain when or if any of the Gates Foundation's condom picks will be ready for prime time. In a follow-up interview with some of the grant winners for *Mic* magazine, they admitted to the challenges inherent in bringing a new product to the market. When Mark McGlothlin of Apex Medical Technologies was asked for updates on his product, the Ultra-Sensitive Reconstituted Collagen Condom, he "seemed almost amused." There were no updates, he said, other than, "It probably is more than a million dollars just to get through FDA approval. It's a brutal process."[12] Until one of these condom innovations emerges from the tangle of the device approval process, we must live with the condoms we have.

Most of us may be unaware, but we have had other barrier protection besides the male condom at our disposal for almost thirty years. After crashing and burning in the United States, one beleaguered protection device was reborn abroad, then feebly resurrected at home, and is now currently living as a recluse. But the time may be right for this product to emerge once again, and if it does, one woman is poised to catapult it between the legs of millions of Americans.

Good for the Goose and the Gander

In the legend of the Phoenix, the mythical bird lives for hundreds of years before preparing itself for death by building a funeral pyre and bursting into flame. After arising from its own ashes, it is reborn into a new life. The poetic metaphor has served as fodder for philosophers, religious figures, and writers to describe recovery from any number of catastrophes: spiritual crises, loss of a loved one, addiction, even divorce.

It was in June of 2008 that Lisa Kinsella, a mother of two living in Chicago, suddenly found herself sitting in a pyre surrounded by flames. That was the month Kinsella and her husband decided to split, and in the same week that she filed her divorce papers, she was diagnosed with breast cancer. Over the next few months, she underwent a double mastectomy and reconstructive surgery. Once her bustline had healed, she turned her attention to the rest of her life. As she thought about having sex again, she realized that she was facing a new world of STIs, a topic she hadn't given much thought to during her married years.

Kinsella still needed contraception, yet hormonal methods such as the Pill, patch, or ring were off limits to her because of her breast cancer history. None of those would give her STI protection anyway, which she would certainly need with a new partner. She wondered why there wasn't an option for STI prevention that she could control on her end. "I felt like there had to be many women in my situation. Where they can't or don't want to take hormones yet still want to be in charge of their own protection." She likened sex to driving, and she wanted to be in control of whatever was keeping her safe behind the wheel: "Why shouldn't I be able to buckle my own seat belt?"

After more than a decade as a sales executive, Kinsella saw

an opportunity to pull an embattled product out of the grave and launch a new career as an entrepreneur. She set her sights on reviving the internal condom, a polyurethane sheath with an inner and outer ring that was designed to be worn inside the vagina, originally marketed in 1993 as the "female condom."

The female condom was certainly a product in need of a do-over. First, there was the matter of its appearance. Its long length (seventeen centimeters) and ample girth gave it a somewhat ungainly look, leaving users with redundant material hanging out after insertion. Upon its initial release, journalists had a field day making fun of its appearance. Sociologist Amy Kaler compiled these jabs in her analysis of the public's response to the product, including comparisons to "a jellyfish, a windsock, a fire hose, a colostomy bag, a Baggie . . . a raincoat for a Slinky toy or a 'contraption used to punish fallen virgins in the Dark Ages.'"[13] In an interview for *Mosaic*, its inventor, Danish physician Lasse Hessel, admitted that its design could be improved. "How can my ugly, clumsy female condom get any worse? It can only get much better."[14]

Then there was the issue of the name. Wisconsin Pharmacal, who bought the initial rights for the device from Hessel, decided to call it the Reality Female Condom, which later became known simply as FC1, in contrast to its eventual successor, the Female Condom 2 (FC2). After her experience in sales, Kinsella took offense at the branding, or lack thereof. "We don't call ourselves 'females,' that's so unsexy. It's clinical." Plus, no one wants to use a product meant for someone else, and the branding inadvertently ostracized a market of potential users: men having sex with men who were using the "female condom" for anal sex.

There were other unexpected problems plaguing the product, including surprise sound effects. Friction from thrusting would cause the device to rustle or squeak. And the squeaking wasn't isolated to penile thrusting. A colleague who studied the internal condom recounted how one user placed it inside his butt before going out for the night. The simple the act of walking caused his butt to squeak for hours: not an effective mating call when one is trying to get laid.

Perhaps people could have gotten over its appearance and squeakiness, but the internal condom was also significantly more expensive than the external condom, costing up to five dollars apiece. While sales flagged in the United States, in 1995, Wisconsin Pharmacal received a call from the Zimbabwe's Ministry of Health and Child Welfare. They had received a petition signed by thirty thousand women demanding that the government bring the internal condom to Zimbabwe.

Although their product didn't catch on with American women, perhaps they would have better luck abroad? Wisconsin Pharmacal packed up their condoms, changed their name to the Female Health Company, and shifted their attention to the public sector abroad. They changed materials from polyurethane (FC1) to a less noisy nitrile (FC2), moved manufacturing to Malaysia to lower costs, and worked with governments and the United Nations Population Fund (UNFPA, formerly the United Nations Fund for Population Activities) to help the FC2 gain traction. By 2010, the UNFPA estimated that fifty million internal condoms had been distributed worldwide.

After gaining FDA approval for the FC2 in March 2009, the Female Health Company attempted to resuscitate it in the United States. According to Emily Anthes in her piece for *Mosaic,* the company launched ad campaigns in major cities, such San Francisco, New York, and Washington, D.C. LGBTQ service organizations also tried to promote it with the anal sex crowd, creating whimsical sleeves to cover the original female-specific packaging, labeled "Manhole Cover" and "Catcher's Mitt."

Meanwhile, Lisa Kinsella was dreaming about bringing internal condoms back to users in the UK and the United States. "This is a neat product. It just had a brand issue and maybe a timing issue, maybe 1992 wasn't the best time or 2009 wasn't the best time." But she saw these problems as fixable. "We're trying to get people to choose a different behavior. It's got to be sexy. It's got to feel good. Women have to be drawn to this, and men too. Of course, the companies have done the opposite. It's been a repellent. Everyone hates condoms. So why are you going to call it a condom when it's technically not a condom?"

After four and a half years of tinkering, she launched a line of internal condoms in the UK in 2016, or as Kinsella preferred to refer to them, "pleasurable protectors" or "a second skin." She chose an empowering yet casual name: LUWI (Let Us Wear It), pronounced *Louie*. Although a release in the United States was her ultimate goal, in 2016, the internal condom was still classified as a Class III device by the FDA (similar to surgical devices like pacemakers or breast implants). Kinsella estimated that obtaining FDA approval for a Class III product would take $1 million and at least a two-year approval process.

Besides, it seemed the internal condom had gone into hiding in the United States. By June 2017, the FC2 was pulled from drugstore shelves and moved behind the counter; it became prescription only.[15] While it was required to be covered under the Affordable Care Act, FC2's manufacturers had thrown in the towel on marketing. According to Kinsella, "They didn't deploy any sales strategy to communicate with physicians or those in a position to write a prescription for it, or to promote it and talk about it." She felt that "the market is ripe for something like LUWI to come in and say, 'Look, there is an option and this option is smart and sexy and it feels better and hey, it's available here [in the United States].'"

Then Kinsella had a stroke of luck. In September 2018, the

FDA announced that it would downgrade the female condom to a Class II device, similar to tampons or standard external condoms. They would also rename it from the "female condom" to the "single-use internal condom" to reflect how it was used for vaginal or anal sex by people of all genders.[16] This downgrading meant that instead of $1 million and two years, it might take only $100,000 and six months to get a new internal condom approved.

Meanwhile, as Kinsella was waiting for the FDA's order to go into effect, her breast cancer recurred. Kinsella decided to start radiation just days after the FDA's order was executed on October 29, 2018. The daily radiation sessions gave her time to "meditate on my next steps, as my energy was being zapped on the other end." Her post-radiation fatigue did not dampen her sense of determination. Now that the FDA had relaxed its classification, she saw a path forward.

By this time, Kinsella was out of money to get LUWI through the FDA, but that hasn't stopped her: "I got another job. I'll be working to self-fund the approval. We've got our prototype, we've got everything we need. So we'll be through the FDA, maybe in the second quarter of 2020." Other competitor products sold abroad, such as the Woman's Condom or the Cupid, may also follow suit. What remains to be seen is whether a critical mass of products will be able to generate interest with a new generation of users.

Regardless of whether the third time's the charm for the internal condom, a movement is under way to put products for women's sexual health in women's hands. Right now, internal and external condoms are the only multipurpose prevention technologies (MPTs) that provide contraception and STI/HIV prevention, but hopefully that won't be true for long. According to Bethany Young Holt, who oversees the global Initiative for Multipurpose Prevention Technologies (IMPT), there are currently twenty MPTs in various stages of development: pills, rings, diaphragms, gels, injectables, and implants. Each product provides protection against at least two conditions: unplanned pregnancy, and HIV or other STIs. One vaginal gel is contending to be the mother of MPTs, potentially protecting against chlamydia, gonorrhea, HIV, HPV, herpes, and unplanned pregnancy.

With any of these potential prevention products, the likelihood of success is low. Young Holt's job is to herd all players that could move one of these MPTs along the development pipeline and keep them talking to one another: funders, researchers, product developers, nonprofit organizations, public health entities. She doesn't really care which products win as long as at least one of them does. "We don't have a particular horse in the race, but we're absolutely bound to Multipurpose Prevention Technologies because we feel passionate that there's promise for these new types of technology and prevention."

But one of the most important factors contributing to an MPT's success is out of Young Holt's hands: money. "It is abundantly clear to me after talking to these researchers and innovators that there has to be more funding. My huge concern is that if money doesn't come in to support innovators, support postdoctoral trainees, etc., they're going to get out of this field and go into diabetes or some other field where there is funding. And we're going to be stuck."

It feels like we've already been stuck for a while. In this era of incredible technologic innovation, why are internal and external condoms still our only prevention option against STIs? As Kaleigh Rogers of *Vice* aptly inquired, "Shouldn't we have some STI-blasting laser by now?"[17] Unfortunately, such a laser is not imminent, and neither is an STI vaccine. As of this writing, there are no forthcoming vaccines against chlamydia, gonorrhea, or syphilis, the three most common bacterial STIs.

Until an STI vaccine falls out of the sky, our options for prevention are limited. Wouldn't it be great if we could just pop a pill that would protect us from STIs, the way that PrEP is used today for HIV? What if this pill had already existed for fifty years and we'd just never thought to use it?

The Siren and the Rocks

In his office at the Los Angeles LGBT Center in Hollywood, medical director Robert Bolan sat mulling over a wicked problem. For the past decade, he'd sat and watched as syphilis and other STIs rose dramatically at home and in the rest of the country, particularly among men who had sex with men. The CDC's annual surveillance report verified what he already knew from his clinical experience: half of the syphilis infections were happening in people already living with HIV. His clinics were seeing ever increasing numbers of patients with STIs, but the local increases weren't his primary concern. He worried that this was a larger problem that was starting to spiral out of control, and he wondered what could be done to stem the tide of syphilis among men who had sex with men.

In some ways, Bolan knew his hands were tied. Clearly, some of these men were serving as core transmitters of syphilis and other STIs in the community. But he wasn't about to tell people to stop having sex. And getting these patients to use condoms would be a hard, if not impossible, nut to crack. From his conversations with patients and from reading studies about "safer sex fatigue," he knew that this concept was real. Men living with HIV were tired of being barraged with safer sex messages, and those who got fed up with playing it safe went back to having sex without condoms. Many HIV patients were on medication and had undetectable viral loads, and others only had sex with other HIV-positive men (a practice known as *serosorting*). For both reasons, they weren't worried about transmitting HIV. Plus, other STIs didn't seem to be as big of a deal compared to HIV. What was a little syphilis in the grand scheme of things?

Bolan and his colleagues decided to try a new tactic. Their HIV-positive patients were already used to taking daily medication. Perhaps they would be also willing to take daily antibiotics to prevent syphilis and other STIs? They decided to try it with doxycycline, an antibiotic that had been around since 1967. It had long been used safely as prophylaxis for malaria and was effective for the treatment of syphilis and chlamydia; this made it a logical candidate to try for STI prophylaxis.

He recruited thirty HIV-positive men who had sex with men, each of whom had been diagnosed and treated for syphilis at least twice since their HIV diagnoses. This was a litmus test of sorts, as syphilis is the least common of the bacterial STIs. So someone who has had syphilis twice is certainly having unprotected sex in high-risk networks and is almost certain to catch other STIs in the future.

Bolan randomly divided the men into two groups. One group would take a single pill of doxycycline daily. To this group, he was clear that "we have no knowledge about whether doxycycline would prevent syphilis and . . . condoms should be consistently used." The other group was paid at each visit if they tested negative for syphilis, gonorrhea, and chlamydia, a practice used in substance abuse treatment called *contingency management*—the longer you stay sober, the greater the rewards. In this case, participants received escalating sums of fifty, seventy-five, and one hundred dollars if they tested negative for STIs at weeks twelve, twenty-four, and thirty-six in the study.

While this would be an effective get-rich-quick-scheme for

some of us, Bolan didn't think that financial incentives would really keep participants STI-free. Bolan had other more practical goals. He wanted to see whether participants assigned to doxycycline would: 1) take the antibiotics, 2) show up for study visits, and 3) start having more unprotected sex because they were taking an STI prophylaxis. But he didn't think he had enough participants to ultimately show an effect of the antibiotics on syphilis or other STIs.

After thirty-six weeks, adding up all the chlamydia, gonorrhea, and syphilis infections, the group taking doxycycline had fewer STIs (six infections) than the group being paid to stay STI-free (fifteen infections).[18] The difference was statistically significant, something Bolan didn't expect with so few participants.

Given that most strategies have failed to curb STIs in men who have sex with men, Bolan's findings were intriguing. In a press release for the study, Bolan was cautiously optimistic about what the results might mean. "We need to conduct a further study with a much larger sample size to determine if we can obtain the same results—if not better results. If our findings are confirmed in a larger study, offering prophylaxis against syphilis . . . could significantly change the dynamics of not only new syphilis infections but possibly also new HIV infections in our community [as syphilis increases the risk of acquiring HIV]. We may be on to something that's unprecedented . . . and very much needed."

Just as Bolan was publishing his findings, Jean-Michel Molina and his colleagues in France were working on their own study of doxycycline to prevent STIs. They already had willing participants, men who had condomless sex with other men, enrolled in an HIV prevention study called IPERGAY (Intervention Préventive de l'Exposition aux Risques avec et pour les Gays). As opposed to Bolan's strategy of taking a doxycycline pill daily, Molina's team had men take doxycycline twenty-four hours *after* unprotected sex, a sort of morning-after pill for STIs.

The results were impressive. After nine months of follow-up, men who took doxycycline after sex were 70 percent less likely to have a new infection with chlamydia or syphilis.[19] Molina didn't find any difference for gonorrhea, which wasn't a surprise— antibiotic resistance to doxycycline has become so high, that class of drugs is no longer effective for gonorrhea treatment.

And therein lies the rub with prophylactic antibiotics. No one knows whether the prophylaxis strategy will simply trade one problem for another: fewer STIs for more antibiotic resistance. And there are other unanswered questions: Would long-term use of antibiotics harm the gut microbiome? Would people taking prophylaxis have more unprotected sex?

When the IPERGAY doxycycline study hit the press, the findings added more fuel to debate sparked by Robert Bolan's study.[20,21] Yes, this antibiotic might work to prevent STIs, but should we use it?

Worldwide, there are five studies under way that will soon help answer this question. But there's already interest among potential users. In a 2018 survey of more than 1,300 users of a gay social networking app, 84 percent said they'd be interested in trying doxycycline for STI prevention. A similar percentage said they'd be willing to enroll in a study to test it out, even if half of them knew that they'd get a placebo.[22]

I ran this concept by Brad Spellberg, an infectious disease specialist and author of *Rising Plague: The Global Threat from Deadly Bacteria and Our Dwindling Arsenal to Fight Them*. In no uncertain terms, he thought using antibiotics this way was a bad idea. "Ever since the first antibiotic became available, people have been lured by the siren's song of their effectiveness onto the rocks of using them for prevention," he said. "This may prevent a few infections in the short term, but it has very bad long-term consequences. The more antibiotics are used, the less effective they become. Over time, the bacteria causing STIs will develop increasing resistance, and not only will the drug stop working for prevention, it will no longer be useful for treatment either."

Spellberg was also concerned that unlike HIV, which is only present in our bodies during an infection, bacteria live on and in our bodies normally. Use of antibiotics could ultimately cause resistance in our bodies' normal bacteria just as they do for infectious bacteria. Plus, we'd excrete these antibiotics into the environment through our urine and feces, which could cause antibiotic resistance among bacteria in the environment as well. Bottom line, he felt that "with few exceptions, attempts to use antibiotics to prevent bacterial infections have almost always ended in regret."

According to Stephanie Cohen, medical director of San Francisco City Clinic, the horse may already be out of the barn. "There are patients who are already using it. We have a very educated, savvy, and motivated community." She was willing to consider prescribing it for certain patients: "If someone came to me requesting doxycycline as postexposure prophylaxis and had a history of recurrent chlamydia or syphilis, I would consider it. But I wouldn't recommend it to everyone. There are still a lot of legitimate unanswered questions about what will happen on a population level with antibiotic resistance. But at least on an individual level, it seems safe and is certainly effective."

Even if doxycycline is eventually used for STI prevention, we can't just hand out antibiotics to every sexually active American. Besides, STIs will never disappear just by popping a simple pill. As long as stigma around STIs still exists, some people will continue to avoid getting tested, getting treated, and telling their partners.

Stopping STIgma

In March 2008 at the ceremony for the American Sexually Transmitted Diseases Association's awards in Chicago, Edward "Ned" Hook III from the University of Alabama–Birmingham approached the podium to receive the Distinguished Career Award. As part of the award ceremony, he was asked to give a speech. He had already delivered hundreds of other speeches in his life, but this one came with one important difference: for the first time in his career, he was told to talk about anything he wanted.

From the get-go, it was clear that Hook was not going to deliver a typical acceptance speech, thanking everyone who had helped in his success. Hook decided to use the bully pulpit to discuss challenges in the field of STI control and share his "opinion about how we can do better." This distinguished Southern gentleman had recently experienced a sexual awakening of his own, so to speak. After spending his whole career thinking about sex in terms of STIs, Hook was abandoning this mentality in favor of a different approach, one focused on sexual health. Sexual health should be seen as "a basic human right," he explained. He implored his colleagues to support a "national campaign based on sexual health," and at that moment, Hook was volunteering to be one of its spokesmen.[23]

The concept of sexual health is old hat for the World Health Organization, which has been discussing and defining the idea since 1975. They describe sexual health as "a state of physical, mental and social well-being in relation to sexuality. It requires a positive and respectful approach to sexuality and sexual relationships, as well as the possibility of having pleasurable and safe sexual experiences, free of coercion, discrimination and violence."

But for researchers who defined their careers by the diseases they study, talking about sexual health felt a little outside the box. Should someone who studies syphilis really be talking about healthy sexuality and pleasure?

Hook himself might have once hesitated, but he felt the country had been banging its head against the wall for too long. "With the amount of time and effort and research that we put into STD control, why are we so much less successful than other developed nations?" He knew that our country's efforts in controlling STIs were rooted in a "far-right, moralist imperative," which stigmatized anyone whose sexual relationships didn't fit into the Christian heteronormative mold: one man and one woman for life. In Hook's estimation, this attitude wasn't doing a thing to drive down STIs, and it was time for the nation to change their tack.

At the time, Hook was chairman of the board of the American Sexual Health Association (ASHA). He approached Lynn Barclay, ASHA's CEO, and said, "We're doing it wrong. We shouldn't be focusing on disease. We should be focusing on health." But there was still something in the way. In his foreword to ASHA's book, *Creating a Sexually Healthy Nation*, Hook boiled it down to Americans' "basic embarrassment we have in acknowledging ourselves (and our parents, friends, and children) as sexual beings." And according to Hook, the time had come to get over it.

In 2010, Hook and Barclay decided to go to Atlanta and talk with their colleagues at the CDC to see if they could get the agency's attention. This wasn't the first time CDC officials had broached the topic. In 2001, David Satcher had released *The Surgeon General's Call to Action to Promote Sexual Health and Responsible Sexual Behavior*. The *Call to Action* was the first formal recognition by the U.S. government of the importance of sexual health to enhance the population's health overall. Unfortunately, ten years later, STIs had exploded and HIV rates were climbing; clearly Americans had been called to action, but not in the way that Satcher had intended.

Despite a few naysayers, Hook and Barclay found allies among CDC leadership who were excited about resurrecting a national discussion about sexual health. In April 2010, the CDC invited them to a national consultation to solicit feedback on a green paper, a draft document they'd created to outline a "positive health-based approach addressing sexual behavior across the lifespan and serving as a potential framework for public health action."[24] Kevin Fenton, director of CDC's National Center for HIV/AIDS, Viral Hepatitis, STD, and TB Prevention, took a distinctly sex-positive approach in his remarks to the group. The concept of sexual health went beyond "the absence of disease, dysfunction, or infirmity"; it was a state of sexual well-being, and he saw the nation's sexual health as an urgent public health matter.

Although many hoped that the green paper from the consultation would become a white paper, or official policy document, it hasn't yet come to pass. But the fact that *sexual health* was coming out of the mouths of CDC officials reflected a broader cultural shift, from "sex=death" in the 1980s, to something more holistic in the post-AIDS era. And once researchers and thought leaders such as Ned Hook had embraced sexual health, there would be no return to the old way of thinking. Hook reflected, "It was the tipping point for me."

Embracing sexual health doesn't discount the realities of the STI epidemic, which unequally afflicts the young, the poor, people of color, and gender and sexual minorities. Maria Trent studies STIs among African American youth in Baltimore, one of the hottest spots for STIs in the country. She's learned to walk a fine line between telling the truth and still empowering young people to take control of their sexual health: "Healthy sexuality is so important for adolescents coming into their sexual identity, to have healthy positive sexual relationships. However, in communities that are facing disparities [in STIs and HIV], we do have to find some way to talk about them without it being completely negative. To tell people the truth so that they can mobilize around that; they can really engage in prevention work that can transform disease rates and their community."

In Baltimore, Trent felt that some teens "see STIs as inevitable. It's so common to have an STI, and most STIs are treatable, that there isn't significant stigma attached to it." According to Trent, "The lack of stigma is a good thing, as long as it doesn't make people less diligent about preventing it. That's the balance we're trying to strike here."

This fatalism about STIs is a double-edged sword: it might reduce stigma, but it also dampens enthusiasm for STI prevention and control. Many people don't realize the connections between STIs and infertility, pregnancy complications, cancer, and increased risk of acquiring HIV. These bad outcomes often don't emerge until decades after the fact. In the short term, most STIs don't kill and are easily treated—thus there is little public outrage around the issue. On an individual level, it's understandable: let me take my antibiotics and pretend this never happened. But this ability to sweep STIs under the rug contributes to the lack of advocates for STI prevention.

There are some hopeful signs of change. Joseph Tucker from the University of North Carolina was lecturing at the London School of Hygiene & Tropical Medicine when he pointed out the lack of advocates in the field of STI, a stark contrast to the field of HIV/AIDS. Researcher Emma Harding-Esch, who was in the audience at the time, took to Twitter to chat about it with her colleagues in Europe. "No one wants to be the face of gonorrhoea," she tweeted. But why did that have to be? What about researchers like her who were proud to be associated with STIs?

Jackie Cassell in the UK tweeted back, "Time for us to be advocates in a beauty contest - #FaceOfAnSTI could be a thing." Their colleague Nicola Low in Switzerland promptly offered herself up as the Face of Chlamydia. Harding-Esch encouraged other researchers who were planning on attending a forthcoming international STI research meeting to stand up proudly as the #FaceOfAnSTI. "If STI researchers won't break down the stigma," she asked, "how can we expect others to?"

A lively Twitter chat soon ensued. I volunteered to be the Face of HPV. Bill Miller in Ohio claimed syphilis. Jane Nicholls of the UK photoshopped herself at a black-tie gala with *Trichomonas vaginalis*. Andrew Lau from Australia posed with a stuffed chlamydia toy attacking him from behind (he studies rectal infections).

At the International Society for STD Research meeting in July 2019, I helped organize a photo booth for researchers to take selfies and group shots posing with extra-large plush STIs against a backdrop that read "#FaceOfAnSTI" and "#StopSTIgma." In the span of three days, hundreds of conference-goers from six continents took to Twitter to display their pride in being associated with their favorite STI. By the end of the meeting, the hashtag #FaceOfAnSTI had almost five hundred thousand Twitter impressions and had reached more than ninety thousand Twitter users.

While #FaceOfAnSTI was a just a small blip in the expanse of the Twitterverse, it is paramount that our tribe of STI enthusiasts continues to do whatever we can to reduce stigma. That includes coming out to everyone about what we really do, even when it might be awkward: to our families, friends, even complete strangers.

Once, I found myself on a plane to Las Vegas next to a buyer from Walmart who quickly shared that she was pro-life, a mother of two teenagers, a Bible study leader, and a Zumba instructor. Then she asked point-blank what I did and why I was traveling that day. I hesitated for a moment. Here I was, a pro-choice mother of two, non-churchgoer on my way to lecture about STIs in the state prison system. How would this sit with my current companion? I took the plunge and told her the truth.

Her face lit up. She had recently chaperoned her Bible study group to a conference with thirty thousand Lutheran teens and they'd attended a popular workshop about sex and STIs. She complained that the session focused on the negative consequences of premarital sex without acknowledging that most people, even Christians, still "try before they buy." To emphasize her point, she glanced around, leaned over and lowered her voice, and confessed, "I had sex before I was married."

While it might seem surprising that a Bible study leader would disclose her premarital sex life within fifteen minutes of meeting, this happens to me all the time. She went on to say that she hoped her kids would have premarital sex: she didn't want either of them spending the rest of their lives having bad sex with their spouses. And she wanted them to be savvy about contraception and STIs. Over the next hour, I answered her litany of questions about both topics before our plane hit the tarmac. I marveled at this mother, who defied my preconceived notions of how religious parents approached the topic of sex and sexual health.

This made me reflect on my parents and their attempt at sex education, which can be summarized by a single sentence: "Don't have sex until marriage or we'll kick you out of the house" (of course I had already had sex by the time I heard this). I don't blame them for taking that approach. Topics of contraception, STIs, and sexual health were irrelevant if one conformed to their traditional norms. Somehow, they ended up with a daughter like me, and I feel for them. *Yes, that's my daughter, the one dressed as a giant condom holding the banana over her head. Her father and I are so proud.*

Now I have a chance to do *the talk* a little differently with my own sons. First, it won't be just one talk, but rather dozens of bite-size discussions on today's complicated sexual landscape: hookup apps, consent, pornography, masturbation, pleasure, STIs, HIV, sexual orientation, gender identity, anatomy, healthy relationships, safety, and more. Half the time that I broach these subjects with my older son, he rolls his eyes or clasps his hands over his ears: "Mom, you are so cringey" (preteen speak for "making him cringe"). His discomfort is a sign of victory; he is actually listening, and some part of my message is getting through.

Given who I am, I will likely take it a bit further. When my sons become teenagers, perhaps I can relive my college days by dressing in costume and doing a repeat of my condom show-and-tell. There's so much ground I'd like to cover with them: materials, fit, proper application, and removal. I can feel them cringing already.

Epilogue

Advice to the Omnivores

I'm jealous of Michael Pollan, author of *In Defense of Food* and *The Omnivore's Dilemma*. It's not because of his prodigious writing skill or the wild success of his books. I envy his ability to distill the morass of conflicting information about what to eat into a simple seven-word manifesto: *Eat food. Not too much. Mostly plants.* While all the details within his books now escape me, his manifesto is burned into my brain and the brains of millions of his readers worldwide. I love to think of him relishing in this accomplishment whilst he massages a huge pile of kale with his hands.

I was optimistic when I started writing this book that I could distill everything I had learned about the world of sex, STIs, and HIV into simple words to live by, à la Michael Pollan. But by the end of it, we've covered so much ground that a single manifesto might not suffice. We discussed the lingering stigma of genital herpes, the stunning success of HPV vaccination, the hazards of pubic hair removal, and the dangers of douching. We learned of epidemic increases in congenital syphilis and the threat of antibiotic-resistant gonorrhea. And I couldn't forget the promise of prophylaxis, for both HIV and STIs. Each topic seemed important; how could I possibly choose one?

Then it became clear. The most important lessons I've learned after years of studying STIs are as follows: 1) catching an STI or HIV isn't necessarily related to how many people you have sex with but rather what sexual networks they happen to be a part of, and 2) we are all going to catch something from someone, someday. Many of us won't realize when it happens, but trust me, it does.

With these lessons in mind, I wanted to craft a directive to help people choose sex partners in a way that minimized the risk and potential for regret. I came back to Pollan's manifesto and tried transposing the first part, *Eat food,* to the field of STIs and sexual health—*Have sex.* That's good. I want everyone reading this book to have as much satisfying, consensual sex as their hearts desire.

I hit a snag when I came to the second part of Pollan's manifesto: *Not too much.* His advice works for eating because your stomach and your brain can usually tell you when you are full and when you've overeaten. You will often vomit if you eat or drink something that is harmful. But while some of us have limits on the number of orgasms we can have in a given time frame, our sex organs possess no limitations in the number of people that we can sleep with. Lust can easily overwhelm the brain as it tries to warn you, "Hey, that person is bad news."

However, the phrase *Not too much* does have some utility to STIs. Limiting sex partners or having monogamous relationships is an effective way to reduce the risk of STIs and HIV, but it is by no means foolproof. It is entirely possible to lose your virginity to the one-and-only love of your life, and simultaneously catch an STI or HIV from that person. Therefore, I can't recommend that everyone sit on their hands and remain virginal until they've found *the one*.

So if limiting partners doesn't guarantee freedom from STIs, we must accept that exposure to STIs is the cost of doing business for all of us who have sex. This applies whether our sexual appetites are large or small, whether our tastes in partners are omnivorous or lean toward a particular gender. The dilemma for all of us lies in trying to balance the risks of STIs/HIV against the many benefits of having sex.

For those of you still actively shopping in the sexual marketplace, here is my attempt at a seven-word manifesto: *Have sex with people that you like.* Some will take offense with this and say, "What about love?" I'm all for having sex with someone that you love. But I can't tell people to do things that I haven't always done myself, and I can't go back and rewrite the past (same goes for these seven words: *Never have sex while you are intoxicated*). If you are in a monogamous relationship, it makes sense to alter the words slightly: *Have sex with someone that you like.* If you happen to be monogamous with someone that you don't like, get out of that relationship if you can. Life is too short.

Does having sex with people that you like prevent STIs or HIV? Does it shield you from regret when you get the diagnosis? No. But in order to like someone, you usually know them enough to find *something* redeeming about them. After delivering countless STI diagnoses, I've seen that people feel more regret when they catch an STI from someone who they can't name or someone who's just not a nice person.

People who like their partners are more apt to disclose an STI to them or help their partner(s) get treated. Breaking the chain of infection benefits both the current parties and future partners too. Obviously, there are caveats. People in "monogamous" relationships can be devastated when an STI enters the picture. Feelings can turn on a dime: *I used to like you, but now I want to kill you.* Sex and relationships are complicated, and STIs make them even more so.

I bounced my manifesto off a colleague who is much more sexually active than I am. His eyes widened, and he clapped me on the shoulder: "That's actually my New Year's resolution!" It felt like I was onto something. I tried my friends, some college students, even my yoga teacher. It seems to work for many types of people who are having sex, so I'm sticking with it. See if it works for you.

I'd be remiss if I left out the third part of Pollan's manifesto: *Mostly plants.* The connection to sex is not as obvious. Certain vegetables might be useful as sex toys, but otherwise, I don't think the third part of his manifesto really applies here. However, if you decide to go with a plant as a sex toy, go organic with that cucumber.

Acknowledgments

When I first embarked on the process of writing and publishing a book, I had no idea what it would actually feel like. Now I know. As Anne Lamott aptly described in *Bird by Bird*, it's like the last few weeks of pregnancy, when you are carrying around thirty extra pounds and feel "hormonally challenged up the yang." And in my case, I gestated this baby for three years. It is a relief to push this book out and have it walking around in the world. For this, I must acknowledge the amazing midwifery skills of my agent, Jessica Papin, and my editor, Sarah Murphy, at Flatiron Books.

Thanks to my cadre of colleagues and friends who provided interviews, read drafts, and provided moral support (sometimes all three at once): Nick Van Wagoner, Christine Johnston, Anna Wald, Nicholas Greenaway, Lawrence Corey, Richard Mancuso, Thomas Gaither, Benjamin Breyer, Jonathan Zenilman, Jeanne Marrazzo, Joel Palefsky, John Potterat, Charles Fann, Bob Kohn, Jessica Frasure-Williams, Levi Downs Jr., Lance Retherford, Jonathan Volk, Kim Koester, David Templeton, Nancy Spencer,

Joe Engelman, Jeff Klausner, Rilene Chew Ng, Matthew Golden, Sheila Lukehart, Christian Faulkenberry-Miranda, Pablo Sanchez, King Holmes, Ralph Katz, Alan Katz, Edward "Ned" Hook III, Lisa Kinsella, Brad Spellberg, Maria Trent, Stephanie Cohen, Bethany Young Holt, Susan Philip, Julie Stoltey, Shannon Garrigan, Melanie Arens, Meghan Ward, Deborah Davis, Antonia Torreblanca, Claudia Borzutzky, Healy Smith, Deanna Fink, Michael Allerton, Sally Slome, Erin Whitney, and Dan Wohlfeiler. Special thanks to Jennifer Zakaras for her astute research assistance throughout the whole process.

Thank you to the mentors who took me under their wings along my career journey: Gail Bolan, Heidi Bauer, Carole Bland, Kim Workowski, Joan Chow, Jeanne Marrazzo, and Jimmy Hara, who instilled his love of syphilis deep within my heart.

She keeps her pom-poms well hidden, but Peggy Orenstein happens to be an incredible cheerleader. I feel lucky and grateful that she is rooting for my team.

Anyone who has written a book will tell you that it's a great way to bring out one's tendency toward mental illness, which struck me at multiple points along this journey. To that end, I am thankful to Laura Mason, Rachel Shaw Heron, and Patricia Hart, who each helped to restore my sanity during the process.

Finally, to my family and support network: James and Young Park, Tony Park and Janey Chan, Ted and Mary Byers, Donald and Ariola Dixon, Dora Hernandez, Araceli Guizar, Chih Pi, my children, Nate and Zane, and, most of all, my husband, Matt Dixon. Thank you so much for babysitting, feeding me, and loving me. I'm sorry that I was grumpy for the past three years.

Notes

1
Killing the Scarlet H

1. Wertheim JO, Smith MD, Smith DM, Scheffler K, Kosakovsky Pond SL. Evolutionary origins of human herpes simplex viruses 1 and 2. *Mol Biol Evol.* 2014 Sep 1;31(9):2356–64.

2. Hutfield DC. History of herpes genitalis. *Br J Vener Dis.* 1966 Dec;42(4): 263–8.

3. Moholy-Nagy S. *The Architecture of Paul Rudolph.* New York: Praeger; 1970:233. http://prudolph.lib.umassd.edu/node/4561.

4. United States Food and Drug Administration. Drugs, devices, and the FDA: part 1: an overview of approval processes for drugs. Available at: https://www.fda.gov/Drugs/DevelopmentApprovalProcess/default.htm. Accessed February 18, 2018.

5. King DH. History, pharmacokinetics, and pharmacology of acyclovir. *J Am Acad Dermatol.* 1988;18(1 Pt 2):176–9.

6. Cuatrecasas P. Drug discovery in jeopardy. *J Clin Invest.* 2006;116(11):2837–2842.

7. Leo J. The new scarlet letter. *Time.* 1982;120(5). Available at: http://content.time.com/time/subscriber/article/0,33009,1715020,00.html. Accessed January 7, 2018.

8. McInnis D. Herpes: Burroughs Wellcome sees its treatment as a way to cure profit ills. *New York Times.* June 19, 1983. Available at: https://www.nytimes.com/1983/06/19/business/herpes-burroughs-wellcome-sees-its-treatment-as-a-way-to-cure-profit-ills.html. Accessed February 14, 2018.

9. McCaffrey K. Drugmakers again boost DTC spending, to $5.6 billion

in 2016. *MM&M*. March 3, 2017. Available at: https://www.mmm-online.com/features/has-dtc-in-the-oncology-space-kept-up-with-science/article/747441/. Accessed March 18, 2018.

10. Donohue J. A History of drug advertising: the evolving roles of consumers and consumer protection. Milbank Q. 2006;84(4):659–99.

11. Mogull SA. Chronology of direct-to-consumer advertising regulation in the United States. *AMWA Journal*. 2008;23(3):106–9.

12. Morris LA, Mazis MB, Brinberg D. Risk disclosures in televised prescription drug advertising to consumers. *J Public Policy Mark*. 1989;8:64–80.

13. Stevenson RW. COMPANY NEWS; Glaxo offers $14 billion for Wellcome. *New York Times*. January 24, 1995. Available at: http://www.nytimes.com/1995/01/24/business/company-news-glaxo-offers-14-billion-for-wellcome.html. Accessed March 18, 2018.

14. Corey L, Reeves WC, Chiang WT, Vontver LA, Remington M, Winter C, Holmes KK. Ineffectiveness of topical ether for the treatment of genital herpes simplex virus infection. *N Engl J Med*. 1978 Aug 3;299(5):237–9.

15. Fleming DT, McQuillan GM, Johnson RE, Nahmias AJ, Aral SO, Lee FK, St Louis ME. Herpes simplex virus type 2 in the United States, 1976 to 1994. *N Engl J Med*. 1997 Oct 16;337(16):1105–11.

16. US Preventive Services Task Force, Bibbins-Domingo K, Grossman DC, Curry SJ, Davidson KW, Epling JW Jr, García FA, Kemper AR, Krist AH, Kurth AE, Landefeld CS, Mangione CM, Phillips WR, Phipps MG, Pignone MP, Silverstein M, Tseng CW. Serologic screening for genital herpes infection: U.S. Preventive Services Task Force recommendation statement. *JAMA*. 2016 Dec 20;316(23):2525–30.

17. Agyemang E, Le QA, Warren T, Magaret AS, Selke S, Johnston C, Jerome KR, Wald A. Performance of commercial enzyme-linked immunoassays for diagnosis of herpes simplex virus-1 and herpes simplex virus-2 infection in a clinical setting. *Sex Transm Dis*. 2017 Dec;44(12):763–7.

18. Kaufman A. The STD that keeps most contestants from competing on "The Bachelor." *New York Post*. February 27, 2018. Available at: https://nypost.com/2018/02/27/the-std-that-keeps-most-contestants-from-competing-on-the-bachelor/. Accessed March 1, 2018.

19. Wald A. Hermeneutics of herpes: the American sexually transmitted dis-

eases association distinguished career award lecture. *Sex Transm Dis.* 2017 Jan;44(1):1–5.

20. Mancuso R. *Asking for a Friend: A True Story of Surviving Herpes and Receiving a Functional Cure from a Vaccine—One You Are Not Allowed to Get.* Noisy Cricket; 2017.

21. Johnston C, Gottlieb SL, Wald A. Status of vaccine research and development of vaccines for herpes simplex virus. *Vaccine.* 2016 Jun 3;34(26):2948–52.

22. Xu CC, Dziegielewski PT, McGaw WT, Seikaly H. Sinonasal undifferentiated carcinoma (SNUC): the Alberta experience and literature review. *J Otolaryngol Head Neck Surg.* 2013 Jan 31;42:2.

23. Mancuso R. Patient zero: the life and work of Dr. Bill Halford. Available at: https://www.youtube.com/watch?v=NOjqjfL7pos&feature=youtu.be.

24. Rational Vaccines. Informed consent to participate in a phase I clinical trial of a therapeutic HSV-2 vaccine. Available at: https://kaiserhealthnews.files .wordpress.com/2017/10/consent-form-for-susan.pdf. Accessed June 25, 2018.

25. PR News Wire. First ever human trial of a live attenuated functioning therapeutic herpes vaccine. October 27, 2016. Available at: https://www.prnewswire .com/news-releases/first-ever-human-trial-of-a-live-attenuated-functioning -therapeutic-herpes-vaccine-300345431.html. Accessed June 25, 2018.

26. Halford WP. Genital herpes meets its match: a live HSV-2 ICP0 - virus vaccine that succeeds where subunit vaccines have failed. Available at: https://kaiserhealthnews.files.wordpress.com/2017/10/halford-perspectives -manuscript-dec-2016.pdf. Accessed June 25, 2018.

27. Taylor M. Offshore human testing of herpes vaccine stokes debate over U.S. safety rules. Kaiser Health News. August 28, 2017. Available at: https:// khn.org/news/offshore-rush-for-herpes-vaccine-roils-debate-over-u-s-safety -rules/. Accessed June 25, 2018.

28. Taylor M. St. Kitts launches probe of herpes vaccine tests on U.S. patients. Kaiser Health News. August 31, 2017. Available at: https://khn.org/news/st -kitts-launches-probe-of-herpes-vaccine-tests-on-u-s-patients/. Accessed June 25, 2018.

29. Taylor M. Desperate quest for herpes cure launched 'rogue' trial. Kaiser Health News. October 19, 2017. Available at: https://khn.org/news/desperate -quest-for-herpes-cure-launched-rogue-trial/. Accessed June 25, 2018.

30. Taylor M. Years before heading offshore, herpes researcher experimented on people in U.S. Kaiser Health News. November 21, 2017. Available at: https://khn .org/news/years-before-heading-offshore-herpes-researcher-experimented-on -people-in-u-s/. Accessed June 25, 2018.

31. Taylor M. FDA launches criminal investigation into unauthorized herpes vaccine research. Kaiser Health News. April 18, 2018. Available at: https:// khn.org/news/fda-launches-criminal-investigation-into-unauthorized-herpes -vaccine-research/. Accessed June 25, 2018.

32. Taylor M. Participants in rogue herpes vaccine research take legal action. Kaiser Health News. March 13, 2018. Available at: https://khn.org/news/participants -in-rogue-herpes-vaccine-research-take-legal-action/. Accessed June 25, 2018.

33. Oseso L, Magaret AS, Jerome KR, Fox J, Wald A. Attitudes and willing-ness to assume risk of experimental therapy to eradicate genital herpes simplex virus infection. *Sex Transm Dis.* 2016 Sep;43(9):566–71.

34. Phipps W, Saracino M, Magaret A, Selke S, Remington M, Huang ML, Warren T, Casper C, Corey L, Wald A. Persistent genital herpes simplex vi-rus-2 shedding years following the first clinical episode. *J Infect Dis.* 2011 Jan 15;203(2):180–7.

35. Tronstein E, Johnston C, Huang ML, Selke S, Magaret A, Warren T, et al. Genital shedding of herpes simplex virus among symptomatic and asymptomatic persons with HSV-2 infection. *JAMA.* 2011;305(14):1441–9.

36. Johnston C, Zhu J, Jing L, Laing KJ, McClurkan CM, Klock A, Diem K, Jin L, Stanaway J, Tronstein E, Kwok WW, Huang ML, Selke S, Fong Y, Magaret A, Koelle DM, Wald A, Corey L. Virologic and immunologic evidence of multifocal genital herpes simplex virus 2 infection. *J Virol.* 2014 May;88(9):4921–31.

37. Xu F, Sternberg MR, Kottiri BJ, McQuillan GM, Lee FK, Nahmias AJ, Berman SM, Markowitz LE. Trends in herpes simplex virus type 1 and type 2 seroprevalence in the United States. *JAMA.* 2006 Aug 23;296(8):964–73.

38. Satterwhite CL, Torrone E, Meites E, Dunne EF, Mahajan R, Ocfemia MC, Su J, Xu F, Weinstock H. Sexually transmitted infections among U.S. women and men: prevalence and incidence estimates, 2008. *Sex Transm Dis.* 2013 Mar;40(3):187–93.

39. McQuillan G, Kruzon-Moran D, Flagg EW, Paulose-Ram R. *Prevalence of Herpes Simplex Virus Type 1 and Type 2 in Persons Aged 14–49: United States,*

2015–2016. NCHS Data Brief, no. 304. Hyattsville, MD: National Center for Health Statistics; 2018.

40. Bernstein DI, Bellamy AR, Hook EW 3rd, Levin MJ, Wald A, Ewell MG, Wolff PA, Deal CD, Heineman TC, Dubin G, Belshe RB. Epidemiology, clinical presentation, and antibody response to primary infection with herpes simplex virus type 1 and type 2 in young women. *Clin Infect Dis.* 2013 Feb;56(3):344–51.

41. Roberts CM, Pfister JR, Spear SJ. Increasing proportion of herpes simplex virus type 1 as a cause of genital herpes infection in college students. *Sex Transm Dis.* 2003 Oct;30(10):797–800.

42. Wald A. Genital HSV-1 infections. *Sex Transm Infect.* 2006 Jun;82(3): 189–90.

43. Berkenwald L. This is about genital herpes. December 7, 2011, updated May 26, 2018. Available at: http://www.scarleteen.com/article/bodies/this_is _about_genital_herpes. Accessed June 25, 2018.

2
Bushwhacked

1. Hitchens C. On the limits of self-improvement, part II. *Vanity Fair.* December 2007. Available at: https://www.vanityfair.com/news/2007/12/hitchens 200712. Accessed November 20, 2017.

2. Sherrow V. *Encyclopedia of Hair: A Cultural History.* Westport, CT: Greenwood Press; 2006.

3. Salafi Islam. The sunan of al-fitrah (ones natural state). Available at: http:// www.salafi-islam.com/aqeedah/belief-in-allah-tawhid/tawhid-al-uloohiyyah /the-sunan-of-al-fitrah-ones-natural-state/. Accessed November 20, 2017.

4. Burke J. Did Renaissance women remove their body hair? Available at: https://renresearch.wordpress.com/2012/12/09/did-renaissance-women -remove-their-body-hair/. Accessed November 20, 2017.

5. Barrow A. Pubic wigs. In: Blakemore C, Jennett S, eds. *The Oxford Companion to the Body.* 2nd ed. Oxford, UK: Oxford University Press; 2001.

6. White R. Girl talk: vajazzling my genital warts made me feel better about having an STI. Available at: http://www.thefrisky.com/2011-04-06/girl-talk -vagazzling-my-genital-warts-made-me-feel-better-about-having/. Accessed November 22, 2017.

7. U.S. Fish and Wildlife Service. Endangered species—listing and critical habitat. Available at: https://www.fws.gov/endangered/what-we-do/listing-overview.html. Accessed November 16, 2017.

8. Herbenick D, Hensel D, Smith NK, Schick V, Reece M, Sanders SA, Fortenberry JD. Pubic hair removal and sexual behavior: findings from a prospective daily diary study of sexually active women in the United States. *J Sex Med.* 2013 Mar;10(3):678–85.

9. Osterberg EC, Gaither TW, Awad MA, Truesdale MD, Allen I, Sutcliffe S, Breyer BN. Correlation between pubic hair grooming and STIs: results from a nationally representative probability sample. *Sex Transm Infect.* 2017 May;93(3):162–6.

10. Eto A, Nakamura M, Ito S, Tanaka M, Furue M. An outbreak of pubic louse infestation on the scalp hair of elderly women. *J Eur Acad Dermatol Venereol.* 2017 Feb;31(2):e79–80.

11. Haddad NM, Brudvig LA, Clobert J, Davies KF, Gonzalez A, Holt RD, Lovejoy TE, Sexton JO, Austin MP, Collins CD, Cook WM, Damschen EI, Ewers RM, Foster BL, Jenkins CN, King AJ, Laurance WF, Levey DJ, Margules CR, Melbourne BA, Nicholls AO, Orrock JL, Song DX, Townshend JR. Habitat fragmentation and its lasting impact on Earth's ecosystems. *Sci Adv.* 2015 Mar 20;1(2):e1500052.

12. Armstrong NR, Wilson JD. Did the "Brazilian" kill the pubic louse? *Sex Transm Infect.* 2006 Jun;82(3):265–6.

13. Dholakia S, Buckler J, Jeans JP, Pillai A, Eagles N, Dholakia S. Pubic lice: an endangered species? *Sex Transm Dis.* 2014 Jun;41(6):388–91.

14. Associated Press. Dutch museum pleads for crab lice donor. Available at: http://www.foxnews.com/story/2007/10/22/dutch-museum-pleads-for-crab-lice-donor.html. Accessed November 19, 2017.

15. Zenilman J. From the boudoir to the bordello: sexually transmitted diseases and travel. In: Schlossberg D, ed. *Infections of Leisure.* 4th ed. Washington, D.C.: ASM Press; 2009.

16. United Nations World Tourism Organization. World Tourism Barometer. Available at: https://www.unwto.org/world-tourism-barometer-n18-january-2020. Accessed June 8, 2020.

17. Reed DL, Light JE, Allen JM, Kirchman JJ. Pair of lice lost or parasites

regained: the evolutionary history of anthropoid primate lice. *BMC Biol.* 2007 Mar 7;5:7.

18. Truesdale MD, Osterberg EC, Gaither TW, Awad MA, Elmer-DeWitt MA, Sutcliffe S, Allen I, Breyer BN. Prevalence of pubic hair grooming-related injuries and identification of high-risk individuals in the United States. *JAMA Dermatol.* 2017 Nov 1;153(11):1114–21.

19. *Transactions of the American Urological Association 5.* Brookline, MA Riverdale Press; 2011:14–15.

20. Gaither TW, Fergus K, Sutcliffe S, Cedars B, Enriquez A, Lee A, Mmonu N, Cohen S, Breyer B. Pubic hair grooming and sexually transmitted infections: a clinic-based cross-sectional survey. *Sex Transm Dis.* 2020 Jun;47(6):419–425.

21. Boxman IL, Hogewoning A, Mulder LH, Bouwes Bavinck JN, ter Schegget J. Detection of human papillomavirus types 6 and 11 in pubic and perianal hair from patients with genital warts. *J Clin Microbiol.* 1999 Jul;37(7):2270–3.

22. Villa L, Varela JA, Otero L, Sánchez C, Junquera ML, Río JS, Vázquez F, Veraldi S, Nazzaro G, Ramoni S. Pubic hair removal and molluscum contagiosum. *Int J STD AIDS.* 2016 Jul;27(8):699–700.

23. Desruelles F, Cunningham SA, Dubois D. Pubic hair removal: a risk factor for "minor" STI such as molluscum contagiosum? *Sex Transm Infect.* 2013 May;89(3):216.

24. DeMaria AL, Flores M, Hirth JM, Berenson AB. Complications related to pubic hair removal. *Am J Obstet Gynecol.* 2014 Jun;210(6):528.e1–5.

25. Butler SM, Smith NK, Collazo E, Caltabiano L, Herbenick D. Pubic hair preferences, reasons for removal, and associated genital symptoms: comparisons between men and women. *J Sex Med.* 2015 Jan;12(1):48–58.

26. Rowen TS, Gaither TW, Awad MA, Osterberg EC, Shindel AW, Breyer BN. Pubic hair grooming prevalence and motivation among women in the United States. *JAMA Dermatol.* 2016 Oct 1;152(10):1106–1113.

27. Gaither TW, Awad MA, Osterberg EC, Rowen TS, Shindel AW, Breyer BN. Prevalence and motivation: pubic hair grooming among men in the United States. *Am J Mens Health.* 2017 May;11(3):620–40.

28. Gaither TW, Truesdale M, Harris CR, Alwaal A, Shindel AW, Allen

IE, Breyer BN. The influence of sexual orientation and sexual role on male grooming-related injuries and infections. *J Sex Med.* 2015 Mar;12(3): 631–40.

3
The Garden of Good and Evil

1. Marrazzo JM, Koutsky LA, Kiviat NB, Kuypers JM, Stine K. Papanicolaou test screening and prevalence of genital human papillomavirus among women who have sex with women. *Am J Public Health.* 2001 Jun;91(6):947–52.

2. Koumans EH, Sternberg M, Bruce C, McQuillan G, Kendrick J, Sutton M, Markowitz LE. The prevalence of bacterial vaginosis in the United States, 2001–2004; associations with symptoms, sexual behaviors, and reproductive health. *Sex Transm Dis.* 2007 Nov;34(11):864–9.

3. Hillier SL, Marrazzo J, Holmes KK. Bacterial vaginosis. In: Holmes KK, Sparling PF, Stamm WE, et al., eds. *Sexually Transmitted Diseases.* 4th ed. New York: McGraw-Hill; 2008:737–68.

4. Cohen CR, Lingappa JR, Baeten JM, Ngayo MO, Spiegel CA, Hong T, Donnell D, Celum C, Kapiga S, Delany S, Bukusi EA. Bacterial vaginosis associated with increased risk of female-to-male HIV-1 transmission: a prospective cohort analysis among African couples. *PLOS Med.* 2012;9(6):e1001251.

5. Gardner HL, Dukes CD. Haemophilus vaginalis vaginitis: a newly defined specific infection previously classified non-specific vaginitis. *Am J Obstet Gynecol.* 1955 May;69(5):962–76.

6. Marrazzo JM, Koutsky LA, Eschenbach DA, Agnew K, Stine K, Hillier SL. Characterization of vaginal flora and bacterial vaginosis in women who have sex with women. *J Infect Dis.* 2002 May 1;185(9):1307–13. Epub 2002 Apr 16.

7. Berger BJ, Kolton S, Zenilman JM, Cummings MC, Feldman J, McCormack WM. Bacterial vaginosis in lesbians: a sexually transmitted disease. *Clin Infect Dis.* 1995 Dec;21(6):1402–5.

8. Fredricks DN, Fiedler TL, Marrazzo JM. Molecular identification of bacteria associated with bacterial vaginosis. *N Engl J Med.* 2005 Nov 3;353(18):1899–911.

9. "My Vag." Available at: https://www.youtube.com/watch?v=z726OPwCnjE. Accessed October 13, 2018.

10. Ravel J, Gajer P, Abdo Z, et al. Vaginal microbiome of reproductive-age women. Proceedings of the National Academy of Sciences of the United States of America. 2011;108:4680–4687.

11. Brooks JP, Buck GA, Chen G, Diao L, Edwards DJ, Fettweis JM, Huzurbazar S, Rakitin A, Satten GA, Smirnova E, Waks Z, Wright ML, Yanover C, Zhou YH. Changes in vaginal community state types reflect major shifts in the microbiome. *Microb Ecol Health Dis.* 2017 Apr 10;28(1):1303265.

12. Srinivasan S, Liu C, Mitchell CM, Fiedler TL, Thomas KK, Agnew KJ, Marrazzo JM, Fredricks DN. Temporal variability of human vaginal bacteria and relationship with bacterial vaginosis. *PLOS ONE.* 2010 Apr 15;5(4):e10197.

13. Gajer P, Brotman RM, Bai G, Sakamoto J, Schütte UM, Zhong X, Koenig SS, Fu L, Ma ZS, Zhou X, Abdo Z, Forney LJ, Ravel J. Temporal dynamics of the human vaginal microbiota. *Sci Transl Med.* 2012 May 2;4(132):132ra52.

14. Srinivasan S, Hoffman NG, Morgan MT, Matsen FA, Fiedler TL, Hall RW, Ross FJ, McCoy CO, Bumgarner R, Marrazzo JM, Fredricks DN. Bacterial communities in women with bacterial vaginosis: high resolution phylogenetic analyses reveal relationships of microbiota to clinical criteria. *PLOS ONE.* 2012;7(6):e37818.

15. Keith L, Stromberg P, Krotoszynski BK, Shah J, Dravnieks A. The odors of the human vagina. *Arch Gynakol.* 1975 Dec 16;220(1):1–10.

16. Harris K and Caskey-Sigety L. *The Medieval Vagina.* South Bend, IN: Snark Publishing; 2014.

17. Sohn A. Charles Knowlton, the father of American birth control. JSTOR Daily. March 21, 2018. Available at: https://daily.jstor.org/charles-knowlton -the-father-of-american-birth-control/. Accessed November 13, 2018.

18. Gannon M. 200-year-old douche found under New York's city hall. *Live Science.* February 21, 2014. Available at: https://www.livescience.com/43583 -vaginal-syringe-ny-city-hall.html.

19. Tone A, ed. *Controlling Reproduction: An American History.* Wilmington, DE: Scholarly Resources; 1997.

20. Martino JL, Vermund SH. Vaginal douching: evidence for risks or benefits to women's health. *Epidemiol Rev.* 2002;24(2):109–24.

21. Meltzer T. Vaginoplasty procedures, complications and aftercare. University

of California San Francisco, Transgender Center of Excellence. Available at: http://transhealth.ucsf.edu/trans?page=guidelines-vaginoplasty. Accessed November 15, 2018.

22. Bradshaw CS, Morton AN, Hocking J, et al. High recurrence rates of bacterial vaginosis over the course of 12 months after oral metronidazole therapy and factors associated with recurrence. *J Infect Dis.* 2006;193(11):1478–86.

23. Amaya-Guio J, Viveros-Carreño DA, Sierra-Barrios EM, Martinez-Velasquez MY, Grillo-Ardila CF. Antibiotic treatment for the sexual partners of women with bacterial vaginosis. Cochrane Database Syst Rev. 2016 Oct 1;10:CD011701.

24. Mehta SD. Systematic review of randomized trials of treatment of male sexual partners for improved bacteria vaginosis outcomes in women. *Sex Transm Dis.* 2012 Oct;39(10):822–30.

25. Bilardi JE, Walker SM, Temple-Smith MJ, McNair RP, Mooney-Somers J, Vodstrcil LA, Bellhouse CE, Fairley CK, Bradshaw CS. Women view key sexual behaviours as the trigger for the onset and recurrence of bacterial vaginosis. *PLOS ONE.* 2017 Mar 9;12(3):e0173637.

26. Liu CM, Hungate BA, Tobian AA, Ravel J, Prodger JL, Serwadda D, Kigozi G, Galiwango RM, Nalugoda F, Keim P, Wawer MJ, Price LB, Gray RH. Penile microbiota and female partner bacterial vaginosis in Rakai, Uganda. *mBio.* 2015 Jun 16;6(3):e00589.

27. Muzny CA, Schwebke JR. Biofilms: an underappreciated mechanism of treatment failure and recurrence in vaginal infections. *Clin Infect Dis.* 2015 Aug 15;61(4):601–6.

28. Cohen CR, Wierzbicki MR, French AL, Morris S, Newmann S, Reno H, Hemmerling A. Randomized trial of lactin-V to prevent recurrence of bacterial vaginosis. *N Engl J Med.* 2020;382(20), 1906–1915.

29. Summers D. For the love of Goop, don't steam your vagina. *Daily Beast.* January 29, 2015. Available at: https://www.thedailybeast.com/for-the-love-of-goop-dont-steam-your-vagina. Accessed November 16, 2018.

30. Gunter J. Dear Gwyneth Paltrow I'm a GYN and your vaginal jade eggs are a bad idea. January 17, 2017. Available at: https://drjengunter.wordpress.com/2017/01/17/dear-gwyneth-paltrow-im-a-gyn-and-your-vaginal-jade-eggs-are-a-bad-idea/. Accessed November 16, 2018.

31. Broddesser-Akner T. How Goop's haters made Gwyneth Paltrow's com-

pany worth $250 million. *New York Times Magazine.* July 25, 2018. Available at: https://www.nytimes.com/2018/07/25/magazine/big-business-gwyneth -paltrow-wellness.html. Accessed October 13, 2018.

32. Kelley Drye & Warren. Eggs-ageration: Goop settles with California district attorneys over misleading health claims. Lexology. September 6, 2018. Available at: https://www.lexology.com/library/detail.aspx?g=3d6f9459-0a6a -4754-9d15-584173b5efb9. Accessed November 16, 2018.

33. Gunter J, Parcak S. Vaginal jade eggs: ancient Chinese practice or modern marketing myth? *Female Pelvic Med Reconstr Surg.* 2018 Oct 25.

4
Warts and all

1. Vilos GA. The history of the Papanicolaou smear and the odyssey of George and Andromache Papanicolaou. *Obstet Gynecol.* 1998 Mar;91(3):479–83.

2. American Society for Clinical Oncology. Cervical cancer, statistics. Available at: https://www.cancer.net/cancer-types/cervical-cancer/statistics. Accessed April 1, 2019.

3. McIntyre P. Finding the viral link: the story of Harald zur Hausen. *Cancer World.* July–August 2005:32–7. Available at: http://www.academia.dk/Blog /wp-content/uploads/harald-zur-hausen.pdf. Accessed April 21, 2019.

4. Chesson HW, Dunne EF, Hariri S, Markowitz LE. The estimated lifetime probability of acquiring human papillomavirus in the United States. *Sex Transm Dis.* 2014 Nov;41(11):660–4.

5. Moscicki AB, Schiffman M, Burchell A, Albero G, Giuliano AR, Goodman MT, Kjaer SK, Palefsky J. Updating the natural history of human papillomavirus and anogenital cancers. *Vaccine.* 2012 Nov 20;30 Suppl 5:F24–33.

6. Collins SI, Mazloomzadeh S, Winter H, Rollason TP, Blomfield P, Young LS, Woodman CB. Proximity of first intercourse to menarche and the risk of human papillomavirus infection: a longitudinal study. *Int J Cancer.* 2005 Apr 10;114(3):498–500.

7. Brown DR, Shew ML, Qadadri B, Neptune N, Vargas M, Tu W, Juliar BE, Breen TE, Fortenberry JD. A longitudinal study of genital human papillomavirus infection in a cohort of closely followed adolescent women. *J Infect Dis.* 2005 Jan 15;191(2):182–92.

8. Dunne EF, Unger ER, Sternberg M, McQuillan G, Swan DC, Patel SS, Markowitz LE. Prevalence of HPV infection among females in the United States. *JAMA*. 2007 Feb 28;297(8):813–9.

9. Sichero L, Giuliano AR, Villa LL. Human papillomavirus and genital disease in men: what we have learned from the HIM Study. *Acta Cytol*. 2019;63(2):109–17.

10. Comella L. 20 years later, how the "Sex and the City" vibrator episode created a lasting buzz. *Forbes*. August 7, 2018. Available at: https://www.forbes.com/sites/lynncomella/2018/08/07/20-years-later-how-the-sex-and-the-city-vibrator-episode-created-a-lasting-buzz/#15a4794649b3. Accessed April 20, 2019.

11. Roden RB, Lowy DR, Schiller JT. Papillomavirus is resistant to desiccation. *J Infect Dis*. 1997 Oct;176(4):1076–9.

12. Anderson TA, Schick V, Herbenick D, Dodge B, Fortenberry JD. A study of human papillomavirus on vaginally inserted sex toys, before and after cleaning, among women who have sex with women and men. *Sex Transm Infect*. 2014 Nov;90(7):529–31.

13. Malagón T, Louvanto K, Wissing M, Burchell AN, Tellier PP, El-Zein M, Coutlée F, Franco EL. Hand-to-genital and genital-to-genital transmission of human papillomaviruses between male and female sexual partners (HITCH): a prospective cohort study. *Lancet Infect Dis*. 2019 Mar;19(3):317–26.

14. Widdice LE, Breland DJ, Jonte J, Farhat S, Ma Y, Leonard AC, Moscicki AB. Human papillomavirus concordance in heterosexual couples. *J Adolesc Health*. 2010 Aug;47(2):151–9.

15. Widdice L, Ma Y, Jonte J, Farhat S, Breland D, Shiboski S, Moscicki AB. Concordance and transmission of human papillomavirus within heterosexual couples observed over short intervals. *J Infect Dis*. 2013 Apr 15;207(8):1286–94.

16. Daley J. The worst jobs in science 2007. *Popular Science*. June 13, 2007. Available at: https://www.popsci.com/scitech/article/2007-06/worst-jobs-science-2007. Accessed May 5, 2019.

17. Speed Weed W. Worst jobs in science: the sequel. *Popular Science*. November 2004. Available at: https://books.google.com/books?id=aAAAAAAMBAJ&pg=PA72&lpg=PA72&dq=popular+science+worst+jobs+in+science+2004+anal&source=bl&ots=IzaHxyTSXs&sig=ACfU3U3xwV4CK8CEETV6mat1XBimbwwtbQ&hl=en&sa=X&ved=2ahUKEwj_vK-

OgYXiAhXWHjQIHXmNAFkQ6AEwCHoECAkQAQ#v=onepage&q
=popular%20science%20worst%20jobs%20in%20science%202004%20
anal&f=false. Accessed May 5, 2019.

18. Palefsky JM, Holly EA, Hogeboom CJ, Berry JM, Jay N, Darragh TM. Anal cytology as a screening tool for anal squamous intraepithelial lesions. *J Acquir Immune Defic Syndr Hum Retrovirol.* 1997 Apr 15;14(5):415–22.

19. Palefsky JM, Holly EA, Ralston ML, Jay N. Prevalence and risk factors for human papillomavirus infection of the anal canal in human immunodeficiency virus (HIV)-positive and HIV-negative homosexual men. *J Infect Dis.* 1998 Feb;177(2):361–7.

20. Machalek DA, Poynten M, Jin F, Fairley CK, Farnsworth A, Garland SM, Hillman RJ, Petoumenos K, Roberts J, Tabrizi SN, Templeton DJ, Grulich AE. Anal human papillomavirus infection and associated neoplastic lesions in men who have sex with men: a systematic review and meta-analysis. *Lancet Oncol.* 2012 May;13(5):487–500.

21. Williams AB, Darragh TM, Vranizan K, Ochia C, Moss AR, Palefsky JM. Anal and cervical human papillomavirus infection and risk of anal and cervical epithelial abnormalities in human immunodeficiency virus-infected women. *Obstet Gynecol.* 1994 Feb;83(2):205–11.

22. Silverberg MJ, Lau B, Justice AC, et al. Risk of anal cancer in HIV-infected and HIV-uninfected individuals in North America. *Clin Infect Dis.* 2012;54:1026–34.

23. Poynten IM, Jin F, Templeton DJ, Law C, Roberts J, Cornall A, Molano LM, Ekman D, McDonald R, Farnsworth A, Garland SM, Fairley CK, Hillman FJ, Grulich AE. Clearance of anal HSIL is inversely related to persistent high-risk HPV—three year follow up results from the Study of Prevention of Anal Cancer (SPANC). Presented at 32nd Annual International Papillomavirus Conference, October 2018, Sydney, Australia.

24. Ryan O'Neal interview. *Piers Morgan Tonight.* June 20, 2011. Available at: https://www.youtube.com/watch?v=ABRQjjF5Mio. Accessed April 21, 2019.

25. Mazziota J. Marcia Cross is sharing her anal cancer story in the hopes of ending the "stigma." *People.* March 27, 2019. Available at: https://people.com/health/marcia-cross-anal-cancer-story-end-stigma/. Accessed April 21, 2019.

26. Jannette Howard, Pink Ribbon Breakfast speech, Sydney, Australia. October 16, 2006. Available at: https://www.youtube.com/watch?v=szXd-iiy2r8. Accessed March 31, 2009.

27. Australian Academy of Science. Interview with Professor Ian Frazer, immunologist. Available at: https://www.science.org.au/learning/general-audience/history/interviews-australian-scientists/professor-ian-frazer-immunologist#12. Accessed April 2, 2019.

28. Cancer Australia. HPV vaccination uptake. Available at: https://ncci.canceraustralia.gov.au/prevention/hpv-vaccination-uptake/hpv-vaccination-uptake. Accessed March 3, 2020.

29. Hall MT, Simms KT, Lew JB, Smith MA, Brotherton JM, Saville M, Frazer IH, Canfell K. The projected timeframe until cervical cancer elimination in Australia: a modelling study. *Lancet Public Health*. 2019 Jan;4(1):e19-e27.

30. Chow EPF. 90% Reduction in genital warts in Australia. Presented at 32nd Annual International Papillomavirus Conference, October 2018, Sydney, Australia. Available at: https://kirby.unsw.edu.au/news/90-decline-genital-warts-young-australians. Accessed May 5, 2019.

31. Burchell AN, Coutlée F, Tellier PP, Hanley J, Franco EL. Genital transmission of human papillomavirus in recently formed heterosexual couples. *J Infect Dis*. 2011;204(11):1723–1729.

32. Centers for Disease Control and Prevention. 2008 through 2017 adolescent HPV vaccination coverage trend report. Available at: https://www.cdc.gov/vaccines/imz-managers/coverage/teenvaxview/data-reports/hpv/trend/index.html. Accessed May 4, 2019.

33. Oliver SE, Unger ER, Lewis R, McDaniel D, Gargano JW, Steinau M, Markowitz LE. Prevalence of human papillomavirus among females after vaccine introduction-National Health and Nutrition Examination Survey, United States, 2003–2014. *J Infect Dis*. 2017 Sep 1;216(5):594–603.

34. McClung NM, Gargano JW, Park IU, et al. Estimated number of cases of high-grade cervical lesions diagnosed among women—United States, 2008 and 2016. *MMWR Morb Mortal Wkly Rep* 2019;68:337–43. http://dx.doi.org/10.15585/mmwr.mm6815a1.

35. Andersson L. RETRACTED: increased incidence of cervical cancer in Sweden: possible link with HPV vaccination. *Indian J Med Ethics*. May 2018.

Available at: https://ijme.in/articles/increased-incidence-of-cervical-cancer-in
-sweden-possible-link-with-hpv-vaccination/?galley=html. Accessed April 1,
2019.

36. Motta M, Callaghan T, Sylvester S. Knowing less but presuming more:
Dunning-Kruger effects and the endorsement of anti-vaccine policy attitudes.
Soc Sci Med. 2018 Aug;211:274–81.

37. Betsch C, Renkewitz F, Betsch T, Ulshöfer C. The Influence of Vaccine-
critical Websites on Perceiving Vaccination Risks. *Journal of Health Psychology*.
2010;15:446–455.

38. Ogilvie GS, Phan F, Pedersen HN, Dobson SR, Naus M, Saewyc EM.
Population-level sexual behaviours in adolescent girls before and after introduction of the human papillomavirus vaccine (2003–2013). *CMAJ*. 2018 Oct
15;190(41):E1221–26.

39. Gee J, Naleway A, Mittendorf K, Irving S, Henninger M, Crane B, Smith
N, Daley M. Incidence of primary ovarian insufficiency following adolescent
vaccination. Presented at 32nd Annual International Papillomavirus Conference, October 2018, Sydney, Australia.

40. Gee J, Arana J. Postural orthostatic tachycardia syndrome after human
papillomavirus vaccination. Presented at 32nd Annual International Papillomavirus Conference, October 2018, Sydney, Australia.

5
Affectionate and Popular

1. Lee KC, Ladizinski B. The clap heard round the world. *Arch Dermatol*.
2012 Feb;148(2):223.

2. Bierman W, Levenson CL. The treatment of gonorrhea arthritis by means
of systemic and additional focal heating. *Am J Med Sci*. 1936;191:55–65.

3. Golden MR, Whittington WL, Handsfield HH, Hughes JP, Stamm WE,
Hogben M, Clark A, Malinski C, Helmers JR, Thomas KK, Holmes KK.
Effect of expedited treatment of sex partners on recurrent or persistent gonor-
rhea or chlamydial infection. *N Engl J Med*. 2005 Feb 17;352(7):676–85.

4. Wolfheiler D, Potterat JJ. How do sexual networks affect HIV/STD prevention? Available at http://caps.ucsf.edu/uploads/pubs/FS/pdf/networksFS
.pdf. Accessed May 20, 2015.

5. Potterat JJ. *Seeking the Positives: A Life Spent on the Cutting Edge of Public Health.* North Charleston, SC: Createspace; 2015.

6. Morris M. Sexual networks and HIV. *AIDS.* 1997;11:S209–16.

7. Centers for Disease Control and Prevention. STDs in racial and ethnic minorities. Available at: http://www.cdc.gov/std/stats14/minorities.htm. Accessed June 20, 2016.

8. Sanders SA, Reece M, Herbenick D, Schick V, Dodge B, Fortenberry JD. Condom use during most recent vaginal intercourse event among a probability sample of adults in the United States. *J Sex Med.* 2010 Oct;7 Suppl 5:362–73.

9. Trujillo L, Chapin-Bardales J, German EJ, Kanny D, Wejnert C. Trends in sexual risk behaviors among Hispanic/Latino Men who have sex with men— 19 urban areas, 2011–2017. *MMWR Morb Mortal Wkly Rep* 2019;68:873–879. http://dx.doi.org/10.15585/mmwr.mm6840a2external icon.

10. Rosenberg ES, Khosropour CM, Sullivan PS. High prevalence of sexual concurrency and concurrent unprotected anal intercourse across racial/ethnic groups among a national, web-based study of men who have sex with men in the United States. *Sex Transm Dis.* 2012;39(10):741–6.

11. Stenger MR, Bauer H, Torrone E, et al. Denominators matter: trends in Neisseria gonorrhoeae incidence among gay, bisexual and other men who have sex with men (GBMSM) in the U.S.—findings from the STD Surveillance Network (SSuN) 2010–2013. *Sex Transm Infect.* 2015;91(Suppl 2):A178–9.

12. de Voux A, Kidd S, Grey JA, Rosenberg ES, Gift TL, Weinstock H, Bernstein KT. State-specific rates of primary and secondary syphilis among men who have sex with men—United States, 2015. *MMWR Morb Mortal Wkly Rep.* 2017 Apr 7;66(13):349–54.

13. TEDxSF-Nicole Daedone-orgasm: the cure for hunger in the Western woman. Available at: https://www.youtube.com/watch?v=s9QVq0EM6g4. Accessed March 2, 2020.

14. Torrone E, Papp J, Weinstock H; Centers for Disease Control and Prevention (CDC). Prevalence of chlamydia trachomatis genital infection among persons aged 14–39 years—United States, 2007–2012. *MMWR Morb Mortal Wkly Rep.* 2014 Sep26; 63(38):834–8.

15. Torrone E. Personal communication. July 7, 2016.

16. Rotello G. *Sexual Ecology: AIDS and the Destiny of Gay Men.* New York: Penguin; 1997.

17. Worobey M, Gemmel M, Teuwen DE, Haselkorn T, Kunstman K, Bunce M, Muyembe JJ, Kabongo JM, Kalengayi RM, Van Marck E, Gilbert MT, Wolinsky SM. Direct evidence of extensive diversity of HIV-1 in Kinshasa by 1960. *Nature.* 2008 Oct 2;455(7213):661–4.

18. Barabási AL. *Linked: How Everything Is Connected to Everything Else and What It Means for Business, Science, and Everyday Life.* New York: Penguin; 2003.

19. Kohn R, Fann C, Bernstein K, Philip S. Discovery of a large sexual network using routine partner services data, San Francisco, 2013. Presented at 2014 Centers for Disease Control and Prevention STD Prevention Conference, Atlanta, Georgia.

6
Knock, Knock, It's the Sex Detectives

1. Potterat JJ. *Seeking the Positives: A Life Spent on the Cutting Edge of Public Health.* North Charleston, SC: CreateSpace, 2015.

2. Potterat JJ, Spencer NE, Muth SQ, eds. *In the Shadow of Venus: Vignettes from the Venereal World.* North Charleston, SC: CreateSpace, 2017.

3. Healy R. The AIDS tracers. *Life.* October 1987:52–5.

4. D'Souza G, Cullen K, Bowie J, Thorpe R, Fakhry C. Differences in oral sexual behaviors by gender, age, and race explain observed differences in prevalence of oral human papillomavirus infection. Liu X, ed. *PLOS ONE.* 2014;9(1):e86023.

5. Bayer R, Toomey KE. Health law and ethics: HIV prevention and the two faces of partner notification. *Am J Public Health.* 1992;82:1158–64.

6. Rutherford GW, Woo JM. Contact tracing and the control of human immunodeficiency virus infection. *JAMA.* 1988; 259:3609–10.

7. Potterat JJ. Contact tracing's price is not its value. *Sex Transm Dis.* 1997 Oct;24(9):519–21.

8. Potterat JJ. Disease intervention specialists as a corps, not corpse. *Sex Transm Dis.* 2008;35(7):703.

9. Potterat JJ. Partner notification for HIV: running out of excuses. *Sex Transm Dis.* 2003 Jan;30(1):89–90.

10. Golden MR, Hogben M, Handsfield HH, St Lawrence JS, Potterat JJ, Holmes KK. Partner notification for HIV and STD in the United States: low coverage for gonorrhea, chlamydial infection, and HIV. *Sex Transm Dis.* 2003 Jun;30(6):490–6.

11. Golden MR, Hogben M, Potterat JJ, Handsfield HH. HIV partner notification in the United States: a national survey of program coverage and outcomes. *Sex Transm Dis.* 2004 Dec;31(12):709–12.

12. Klausner J. Tracking a syphilis outbreak through cyberspace. In: Dworkin M, ed. *Outbreak Investigations Around the World: Case Studies in Infectious Disease Field Epidemiology.* Burlington, MA: Jones and Bartlett; 2010.

13. Klausner JD, Wolf W, Fischer-Ponce L, Zolt I, Katz MH. Tracing a syphilis outbreak through cyberspace. *JAMA.* 2000;284(4):447–9.

14. Nieves E. Privacy questions raised in cases of syphilis linked to chat room. *New York Times.* August 25, 1999:A1.

15. Torrone, E. Personal communication, September 4, 2020.

16. Pew Research Center. Mobile fact sheet. Available at: http://www.pewinter net.org/fact-sheet/mobile/. Accessed July 8, 2018.

17. Smith C. 50 interesting Tinder statistics and facts (September 2018) | by the numbers. Available at: https://expandedramblings.com/index.php/tinder -statistics/. Accessed September 14, 2018.

18. Johnson K. What is Grindr? Everything you need to know about the gay and bisexual dating app. Available at: https://www.leeds-live.co.uk/news/uk -world-news/grindr-app-dating-gay-bisexual-14662617. Accessed September 14, 2018.

19. Kachur R, Strona FV, Kinsey J, Collins D. *Introducing Technology Into Partner Services: A Toolkit for Programs.* Atlanta, GA: Centers for Disease Control and Prevention; 2015.

20. Katz D, Goyette M, Fredericksen R, Wohlfeiler D, Hecht J, Kachur R, Strona F. Acceptability of HIV/STD partner notification using geosocial networking apps. Presented at 2018 STD Prevention Conference, Washington, D.C. Abstract POS 295.

21. Chew Ng R, Nguyen R, Kohn R, Shaw R, Sachdev D, Cohen S, Philip S. Reframing the value of syphilis partner services—San Francisco

2017. Presented at: 2018 STD Prevention Conference, Washington, DC. Abstract 6C2.

22. Ferreira A, Young T, Mathews C, Zunza M, Low N. Strategies for partner notification for sexually transmitted infections, including HIV. Cochrane Database Syst Rev. 2013 Oct 3;(10):CD002843.

7
A Pox on Both Your Houses

1. Frith J. Syphilis—its early history and treatment until penicillin and the debate on its origins. *J Mil Veterans Health*. 2012; 20(4):49–59.

2. Arrizabalaga J, Henderson J, French R. *The Great Pox: The French Disease in Renaissance Europe*. New Haven, CT: Yale University Press; 2014.

3. Silverman ME, Murray TJ, Bryan CS, eds. *The Quotable Osler*. Philadelphia: American College of Physicians; 2008.

4. Osler W. The campaign against venereal disease: Sir William Osler's oration. *Br Med J*. 1917;1(2943):694–6.

5. Brandt AM. *No Magic Bullet: A Social History of Venereal Disease in the United States Since 1880*. New York: Oxford University Press; 1985.

6. Ward D. The geography of the *Ladies' Home Journal*—an analysis of a magazine's audience, 1911–1955. *J Hist*. 2008;34(1):2–14.

7. Parran T. *Shadow on the Land*. New York: Reynal and Hitchcock; 1937.

8. Livingood CS. History of the American Board of Dermatology, Inc. (1932–1982). *J Am Acad Dermatol*. 1982 Dec;7(6):821–50.

9. Nelson N. The civilian education program in the control of syphilis. *JAMA*. 1936;107(11):872–4.

10. Keating P. *Bluff Your Way in Doctoring*. Horsham, UK: Ravette; 1993.

11. Wear J, Holmes KK. *How to Have Intercourse Without Getting Screwed*. Seattle, WA: Madrona; 1976.

12. Centers for Disease Control and Prevention. Primary and secondary syphilis—United States 2000–2001. MMWR. https://www.cdc.gov/mmwr/preview/mmwrhtml/mm5143a4.htm.

13. Centers for Disease Control and Prevention. 2018 STD surveillance: syphilis. Available at: https://www.cdc.gov/std/stats18/syphilis.htm. Accessed June 8, 2020.

14. Williams LA, Klausner JD, Whittington WL, Handsfield HH, Celum C, Holmes KK. Elimination and reintroduction of primary and secondary syphilis. *Am J Public Health.* 1999 Jul;89(7):1093–7.

15. Gage SH. Dark-field microscopy and the history of its development. *Trans Am Microsc Soc.* 1920 Apr;39(2):95–141. Available at: https://www.jstor.org/stable/3221838. Accessed January 1, 2019.

16. Dowell D, Polgreen PM, Beekmann SE, Workowski KA, Berman SM, Peterman TA. Dilemmas in the management of syphilis: a survey of infectious diseases experts. *Clin Infect Dis.* 2009 Nov 15;49(10):1526–9.

17. Turner TB, Hardy PH, Newman B. Infectivity tests in syphilis. *Br J Vener Dis.* 1969 Sep;45(3):183–95. PubMed PMID: 4899592; PubMed Central PMCID: PMC1048462.

18. Lukehart SA. Biology of treponemes. In Holmes KK et al., eds. *Sexually Transmitted Diseases.* New York: McGraw-Hill; 2008.

19. Hayden D. *Pox: Genius, Madness, and the Mysteries of Syphilis.* New York: Basic Books; 2003.

20. Pathela P, Braunstein SL, Blank S, Shepard C, Schillinger JA. The high risk of an HIV diagnosis following a diagnosis of syphilis: a population-level analysis of New York City men. *Clin Infect Dis.* 2015 Jul 15;61(2):281–7.

21. Rodriguez R. Fresno County 2017 crop values rebound, rising to $7 billion. *Fresno Bee.* August 21, 2018. Available at: https://www.fresnobee.com/latest-news/article217097230.html. Accessed January 28, 2019.

22. American Fact Finder. Community facts—San Francisco, California. Available at: https://factfinder.census.gov/faces/nav/jsf/pages/community_facts.xhtml?src=bkmk. Accessed February 3, 2019.

23. California Department of Public Health, STD Control Branch. California STD data tables 2013–2017. Available at: https://www.cdph.ca.gov/Programs/CID/DCDC/CDPH%20Document%20Library/CA-STD-2017-Data-Tables.pdf. Accessed February 20, 2019.

24. Jones JH. *Bad Blood: The Tuskegee Syphilis Experiment.* New York: Free Press; 1993.

25. Heller J. Syphilis victims in U.S. study went untreated for 40 years. *New York Times.* July 26, 1972. Available at: https://www.nytimes.com/1972/07/26/archives/syphilis-victims-in-us-study-went-untreated-for-40-years-syphilis.html. Accessed January 18, 2019.

26. Katz RV, Warren RC, eds. *The Search for the Legacy of the USPHS Syphilis Study at Tuskegee.* Lanham, MD: Lexington Books; 2011.

27. Katz RV, Kegeles SS, Green BL, Kressin NR, James SA, Claudio C. The Tuskegee Legacy Project: history, preliminary scientific findings, and unanticipated societal benefits. *Dent Clin North Am.* 2003 Jan;47(1):1–19.

28. White House Office of the Press Secretary. Remarks by the president in apology for study done in Tuskegee. May 16, 1997. Available at: https://www.cdc.gov/tuskegee/clintonp.htm. Accessed January 16, 2019.

29. Katz RV, Green BL, Kressin NR, James SA, Wang MQ, Claudio C, Russell SL. Exploring the "legacy" of the Tuskegee Syphilis Study: a follow-up study from the Tuskegee Legacy Project. *J Natl Med Assoc.* 2009 Feb;101(2):179–83.

30. Katz RV, Kegeles SS, Kressin NR, Green BL, James SA, Wang MQ, Russell SL, Claudio C. Awareness of the Tuskegee Syphilis Study and the US presidential apology and their influence on minority participation in biomedical research. *Am J Public Health.* 2008 Jun;98(6):1137–42.

31. Centers for Disease Control and Prevention. U.S. Public Health Service Syphilis Study at Tuskegee—frequently asked questions. Available at: https://www.cdc.gov/tuskegee/faq.htm Accessed February 25, 2019.

8
The Path of Least Resistance

1. Wuebker E. Venereal disease visual history archive. Available at: https://vdarchive.newmedialab.cuny.edu/items/browse?collection=9. Accessed March 2, 2019.

2. Rasnake MS, Conger NG, McAllister K, Holmes KK, Tramont EC. History of U.S. military contributions to the study of sexually transmitted diseases. *Mil Med.* 2005 Apr;170(4 Suppl):61–5. PubMed PMID: 15916284.

3. Cutler JC. Gonorrheal Experiment #4. National Archives, records of Dr. John C. Cutler. Available at: https://nara-media-001.s3.amazonaws.com/arcmedia/research/health/cdc-cutler-records/folder-33-gonorrheal-experiment.pdf. Accessed March 18, 2019.

4. Cutler JC. Experimental studies in gonorrhea. Report. National Archives, records of Dr. John C. Cutler. Available at: https://nara-media-001.s3.amazonaws .com/arcmedia/research/health/cdc-cutler-records/folder-15-experimental -studies-in-gonorrhea.pdf. Accessed March 18, 2019.

5. Semeniuk I, Reverby S. A shocking discovery. *Nature*. 2010 Oct 7;467 (7316):645.

6. Presidential Commission for the Study of Bioethical Issues. *Ethically Impossible: STD Research in Guatemala from 1946 to 1948*. Washington, D.C.; 2011. Available at: https://bioethicsarchive.georgetown.edu/pcsbi/sites/default /files/Ethically%20Impossible%20(with%20linked%20historical%20docu ments)%202.7.13.pdf. Accessed March 2, 2019.

7. Zenilman J. The Guatemala sexually transmitted disease studies: what happened. *Sex Transm Dis*. 2013 Apr;40(4):277–9.

8. Castillo M. U.S. Rejects Guatemalans' STD lawsuit, offers aid. CNN. January 10, 2012. Available at: https://www.cnn.com/2012/01/10/world/americas /us-guatemala-std-experiments/index.html. Accessed March 2, 2019.

9. United States Department of State. Tort claims against the U.S. Department of State. Available at: https://www.state.gov/s/l/3202.htm. Accessed March 2, 2019.

10. Stempel J. Johns Hopkins, Bristol-Myers must face $1B in syphilis infections lawsuit. Reuters Health Information. Available at: https://www.reuters .com/article/us-maryland-lawsuit-infections/johns-hopkins-bristol-myers-must -face-1-billion-syphilis-infections-suit-idUSKCN1OY1N3. Accessed January 19, 2019.

11. Philippines Statistics Authority. 2010 population of Olongapo City is twice its population forty years ago. July 10, 2013. Available at: https://psa.gov .ph/content/2010-population-olongapo-city-twice-its-population-forty-years -ago-results-2010-census. Accessed March 19, 2019.

12. Hawaii Tourism Authority. News release HTA: Hawaii visitor statistics released for 2018. Available at: https://governor.hawaii.gov/newsroom/latest -news/news-release-hta-hawaii-visitor-statistics-released-for-2018/. Accessed March 29, 2019.

13. Centers for Disease Control and Prevention. STD success stories: battling antibiotic-resistant gonorrhea: a timeline of coordinated teamwork. Available

at: https://www.cdc.gov/std/products/success/Hawaii-SUCCESS-STORIES
.pdf. Accessed March 21, 2019.

14. Unemo M, Shafer WM. Antimicrobial resistance in Neisseria gonorrhoeae
in the 21st century: past, evolution, and future. *CMR*. Jun 2014;27(3):587–
613.

15. Elwell LP, Roberts M, Mayer LW, Falkow S. Plasmid-mediated beta-
lactamase production in Neisseria gonorrhoeae. *Antimicrob Agents Chemother*.
1977 Mar;11(3):528–33.

16. Barbee LA. Preparing for an era of untreatable gonorrhea. *Curr Opin Infect
Dis*. 2014 Jun;27(3):282–7.

17. Habel, Melissa A., Jami S. Leichliter, Patricia J. Dittus, Ian H. Spicknall,
and Sevgi O. Aral. 2018. "Heterosexual Anal and Oral Sex in Adolescents
and Adults in the United States, 2011–2015." *Sexually Transmitted Diseases*
45 (12): 775–782.

18. Chow EP, Howden BP, Walker S, Lee D, Bradshaw CS, Chen MY, Snow
A, Cook S, Fehler G, Fairley CK. Antiseptic mouthwash against pharyngeal
Neisseria gonorrhoeae: a randomised controlled trial and an in vitro study. *Sex
Transm Infect*. 2017 Mar;93(2):88–93.

19. Emen8. Have we found a new weapon in the fight against gonorrhea? Available at: https://emen8.com.au/health/fitness_and_body/have-we-found-a-new
-weapon-in-the-fight-against-gonorrhea/. Accessed March 29, 2019.

20. Sentencing Project. Criminal justice facts. Available at: https://www
.sentencingproject.org/criminal-justice-facts/. Accessed March 29, 2019.

21. Thomas JC, Sampson LA. High rates of incarceration as a social force associated with community rates of sexually transmitted infection. *J Infect Dis*.
2005 Feb 1;191 Suppl 1:S55–60.

22. Aral SO, Hughes JP, Stoner B, et al. Sexual mixing patterns in the spread of
gonococcal and chlamydial infections. *Am J Public Health*. 1999;89(6):825–833.

23. Andrasik MP, Nguyen HV, George WH, Kajumulo KF. Sexual decision
making in the absence of choice: the African American female dating experience. *J Health Dispar Res Pract*. 2014 Winter;7(7):66–86.

24. Dauria EF, Elifson K, Arriola KJ, Wingood G, Cooper HL. Male incarceration rates and rates of sexually transmitted infections: results from

a longitudinal analysis in a southeastern US city. *Sex Transm Dis.* 2015 Jun;42(6):324–8.

25. Dauria EF, Oakley L, Arriola KJ, Elifson K, Wingood G, Cooper HL. Collateral consequences: implications of male incarceration rates, imbalanced sex ratios and partner availability for heterosexual Black women. *Cult Health Sex.* 2015;17(10):1190–206.

26. Centers for Disease Control and Prevention. STDs in racial and ethnic minorities. Available at: https://www.cdc.gov/std/stats17/minorities.htm. Accessed March 28, 2019.

27. Potterat JJ, Rothenberg RB, Woodhouse DE, Muth JB, Pratts CI, Fogle JS 2nd. Gonorrhea as a social disease. *Sex Transm Dis.* 1985 Jan-Mar; 12(1):25–32.

28. Potterat JJ. "Socio-geographic" space and sexually transmissible diseases in the 1990s. *Today's Life Science.* 1992; 4 (12); 16–22, 31.

9
PrEPared

1. Enovid advertisement. *Can Medical J.* 17 February 1962;353.

2. Eig J. *The Birth of the Pill: How Four Crusaders Reinvented Sex and Launched a Revolution.* New York: W. W. Norton; 2014.

3. The pill: how is it affecting U.S. morals, family life? *U.S. News & World Report.* July 11, 1966.

4. Myers JE, Sepkowitz KA. A pill for HIV prevention: déjà vu all over again? *Clin Infect Dis.* 2013 Jun;56(11):1604–12. doi: 10.1093/cid/cit085. Epub 2013 Feb 13.

5. Grant RM, Lama JR, Anderson PL, McMahan V, Liu AY, Vargas L, Goicochea P, Casapía M, Guanira-Carranza JV, Ramirez-Cardich ME, Montoya-Herrera O, Fernández T, Veloso VG, Buchbinder SP, Chariyalertsak S, Schechter M, Bekker LG, Mayer KH, Kallás EG, Amico KR, Mulligan K, Bushman LR, Hance RJ, Ganoza C, Defechereux P, Postle B, Wang F, McConnell JJ, Zheng JH, Lee J, Rooney JF, Jaffe HS, Martinez AI, Burns DN, Glidden DV; iPrEx Study Team. Preexposure chemoprophylaxis for HIV prevention in men who have sex with men. *N Engl J Med.* 2010 Dec 30;363(27):2587–99.

6. Baeten JM, Donnell D, Ndase P, Mugo NR, Campbell JD, Wangisi J, Tappero JW, Bukusi EA, Cohen CR, Katabira E, Ronald A, Tumwesigye E, Were E, Fife KH, Kiarie J, Farquhar C, John-Stewart G, Kakia A, Odoyo J, Mucunguzi A, Nakku-Joloba E, Twesigye R, Ngure K, Apaka C, Tamooh H, Gabona F, Mujugira A, Panteleeff D, Thomas KK, Kidoguchi L, Krows M, Revall J, Morrison S, Haugen H, Emmanuel-Ogier M, Ondrejcek L, Coombs RW, Frenkel L, Hendrix C, Bumpus NN, Bangsberg D, Haberer JE, Stevens WS, Lingappa JR, Celum C; Partners PrEP Study Team. Antiretroviral prophylaxis for HIV prevention in heterosexual men and women. *N Engl J Med*. 2012 Aug 2;367(5):399–410.

7. Duran D. Truvada whores? *Huffington Post*. November 12, 2012. Updated February 2, 2016. Accessed September 10, 2017. Available at: https://www.huffpost.com/entry/truvada-whores_b_2113588.

8. U.S. Public Health Service. Preexposure prophylaxis for the prevention of HIV infection the United States—2014. Available at: https://www.cdc.gov/hiv/pdf/prepguidelines2014.pdf. Accessed September 1, 2017.

9. Smith DK, Van Handel M, Wolitski RJ, Stryker JE, Hall HI, Prejean J, Koenig LJ, Valleroy LA. Vital signs: estimated percentages and numbers of adults with indications for preexposure prophylaxis to prevent HIV acquisition—United States, 2015. *MMWR Morb Mortal Wkly Rep*. 2015 Nov 27;64(46):1291–5.

10. AIDS Healthcare Foundation. PrEP: the revolution that didn't happen. Available at: https://www.aidshealth.org/#/archives/24286. Accessed June 8, 2017.

11. Centers for Disease Control and Prevention. HIV among gay and bisexual men. https://www.cdc.gov/hiv/group/msm/index.html. Accessed March 7, 2020.

12. Marrazzo JM, Ramjee G, Richardson BA, Gomez K, Mgodi N, Nair G, Palanee T, Nakabiito C, van der Straten A, Noguchi L, Hendrix CW, Dai JY, Ganesh S, Mkhize B, Taljaard M, Parikh UM, Piper J, Mâsse B, Grossman C, Rooney J, Schwartz JL,Watts H, Marzinke MA, Hillier SL, McGowan IM, Chirenje ZM; VOICE Study Team. Tenofovir-based preexposure prophylaxis for HIV infection among African women. *N Engl J Med*. 2015 Feb 5;372(6):509–18.

13. van der Straten A, Stadler J, Montgomery E, Hartmann M, Magazi B, Mathebula F, Schwartz K, Laborde N, Soto-Torres L. Women's experiences

with oral and vaginal pre-exposure prophylaxis: the VOICE-C qualitative study in Johannesburg, South Africa. *PLOS ONE*. 2014 Feb 21;9(2):e89118.

14. van der Straten A, Montgomery ET, Musara P, Etima J, Naidoo S, Laborde N, Hartmann M, Levy L, Bennie T, Cheng H, Piper J, Grossman CI, Marrazzo J, Mensch B; Microbicide Trials Network-003D Study Team. Disclosure of pharmacokinetic drug results to understand nonadherence. *AIDS*. 2015 Oct 23;29(16):2161–71.

15. Centers for Disease Control and Prevention. HIV risk and prevention estimates. Available at: https://www.cdc.gov/hiv/risk/estimates/riskbehaviors .html. Accessed August 30, 2017.

16. Cohen MS, Chen YQ, McCauley M, Gamble T, Hosseinipour MC, Kumarasamy N, Hakim JG, Kumwenda J, Grinsztejn B, Pilotto JH, Godbole SV, Mehendale S, Chariyalertsak S, Santos BR, Mayer KH, Hoffman IF, Eshleman SH, Piwowar-Manning E, Wang L, Makhema J, Mills LA, de Bruyn G, Sanne I, Eron J, Gallant J, Havlir D, Swindells S, Ribaudo H, Elharrar V, Burns D, Taha TE, Nielsen-Saines K, Celentano D, Essex M, Fleming TR; HPTN 052 Study Team. Prevention of HIV-1 infection with early antiretroviral therapy. *N Engl J Med*. 2011 Aug 11;365(6):493–505.

17. Cohen MS, Chen YQ, McCauley M, Gamble T, Hosseinipour MC, Kumarasamy N, Hakim JG, Kumwenda J, Grinsztejn B, Pilotto JH, Godbole SV, Chariyalertsak S, Santos BR, Mayer KH, Hoffman IF, Eshleman SH, Piwowar-Manning E, Cottle L, Zhang XC, Makhema J, Mills LA, Panchia R, Faesen S, Eron J, Gallant J, Havlir D, Swindells S, Elharrar V, Burns D, Taha TE, Nielsen-Saines K, Celentano DD, Essex M, Hudelson SE, Redd AD, Fleming TR; HPTN 052 Study Team. Antiretroviral therapy for the prevention of hiv-1 transmission. *N Engl J Med*. 2016 Sep 1;375(9):830–9.

18. Bavinton BR, Pinto AN, Phanuphak N, Grinsztejn B, Prestage GP, Zablotska-Manos IB, Jin F, Fairley CK, Moore R, Roth N, Bloch M, Pell C, McNulty AM, Baker D, Hoy J, Tee BK, Templeton DJ, Cooper DA, Emery S, Kelleher A, Grulich AE; Opposites Attract Study Group. Viral suppression and HIV transmission in serodiscordant male couples: an international, prospective, observational, cohort study. *Lancet HIV*. 2018 Aug;5(8):e438–47.

19. Rodger AJ, Cambiano V, Bruun T, Vernazza P, Collins S, van Lunzen J, Corbelli GM, Estrada V, Geretti AM, Beloukas A, Asboe D, Viciana P, Gutiérrez F, Clotet B, Pradier C, Gerstoft J, Weber R, Westling K, Wandeler G, Prins JM, Rieger A, Stoeckle M, Kümmerle T, Bini T, Ammassari A,

Gilson R, Krznaric I, Ristola M, Zangerle R, Handberg P, Antela A, Allan S, Phillips AN, Lundgren J; PARTNER Study Group. Sexual activity without condoms and risk of HIV transmission in serodifferent couples when the hiv-positive partner is using suppressive antiretroviral therapy. *JAMA*. 2016 Jul 12;316(2):171–81. doi: 10.1001/jama.2016.5148. Erratum in: *JAMA*. 2016 Aug 9;316(6):667. JAMA. 2016 Nov 15;316(19):2048.

20. Rodger AJ, Cambiano V, Bruun T, Vernazza P, Collins S, Degen O, Corbelli GM, Estrada V, Geretti AM, Beloukas A, Raben D, Coll P, Antinori A, Nwokolo N, Rieger A, Prins JM, Blaxhult A, Weber R, Van Eeden A, Brockmeyer NH, Clarke A, Del Romero Guerrero J, Raffi F, Bogner JR, Wandeler G, Gerstoft J, Gutiérrez F, Brinkman K, Kitchen M, Ostergaard L, Leon A, Ristola M, Jessen H, Stellbrink HJ, Phillips AN, Lundgren J; PARTNER Study Group. Risk of HIV transmission through condomless sex in serodifferent gay couples with the HIV-positive partner taking suppressive antiretroviral therapy (PARTNER): final results of a multicentre, prospective, observational study. *Lancet*. 2019 Jun 15;393(10189):2428–38.

21. Finlayson T, Cha S, Xia M, et al. Changes in HIV Preexposure prophylaxis awareness and use among men who have sex with men—20 urban areas, 2014 and 2017. *MMWR Morb Mortal Wkly Rep*. 2019;68:597–603.

10
When the Rubber Meets the Road

1. Farrington EM, Bell DC, DiBacco AE. Reasons people give for using (or not using) condoms. *AIDS Behav*. 2016 Dec;20(12):2850–62.

2. Daniels K, Abma JC. Current contraceptive status among women aged 15–49: United States, 2015–2017. Available at: https://www.cdc.gov/nchs/data/databriefs/db327-h.pdf. Accessed May 22, 2019.

3. Kann L, McManus T, Harris WA, et al. Youth Risk Behavior Surveillance—United States, 2017. *MMWR Surveill Summ*. 2018;67(No. SS-8):1–114.

4. United Nations, Department of Economic and Social Affairs, Population Division (2015). Trends in Contraceptive Use Worldwide 2015 (ST/ESA/SER.A/349).

5. Yoshida H, Sakamoto H, Leslie A, Takahashi O, Tsuboi S, Kitamura K. Contraception in Japan: current trends. *Contraception*. 2016 Jun;93(6):475–7.

6. Gardiner D. Japan ok's pill, cautiously. September 6, 1999. Available at:

https://www.wired.com/1999/09/japan-oks-pill-cautiously/. Accessed May 22, 2019.

7. Zion Market Research. U.S. condom market size to reach USD 1,680.22 Mn by 2022. Global News Wire. January 10, 2018. Available at: https://www.globenewswire.com/news-release/2018/01/10/1286656/0/en/U-S-Condom-Market-Size-to-Reach-USD-1-680-22-Mn-by-2022-Zion-Market-Research.html. Accessed May 15, 2019.

8. Bill & Melinda Gates Foundation. Global Grand Challenges. Develop the next generation of condom. Available at: https://gcgh.grandchallenges.org/challenge/develop-next-generation-condom-round-12. Accessed May 15, 2019.

9. Salif Sow P, Ward S. Reinventing the condom. *Impatient Optimists*. Available at: https://www.impatientoptimists.org/Posts/2013/03/Reinventing-The-Condom#.XOXnHfZFw6Y. Accessed May 22, 2019.

10. Harrington E. Origami con man. *Washington Beacon*. May 1, 2014. Available at: https://freebeacon.com/issues/origami-con-man/. Accessed May 30, 2019.

11. Harrington E. Origami Condom inventor has to pay back taxpayer funds. *Washington Beacon*. May 1, 2014. https://freebeacon.com/issues/origami-condom-inventor-has-to-pay-back-taxpayer-funds/. Accessed May 30, 2019.

12. Taylor J. 2 years later, here's what happened to Bill Gates' condoms of the future. *Mic*. November 18, 2015. Available at: https://www.mic.com/articles/128850/bill-and-melinda-gates-foundation-condom-contest-where-are-they-now. Accessed June 1, 2019.

13. Kaler A. The female condom in North America: selling the technology of 'empowerment.' *J Gend Stud*. 2004;13(2):139–52. doi: 10.1080/0958923042000217819.

14. Anthes E. The future of sex? *Mosaic*. March 4, 2014. Available at: https://mosaicscience.com/story/future-sex/. Accessed May 30, 2019.

15. Guerrero D. Internal condoms become prescription only. *Plus*. June 6, 2017. https://www.hivplusmag.com/prevention/2017/6/16/internal-condoms-become-prescription-only. Accessed June 1, 2019.

16. Food and Drug Administration. Obstetrical and gynecological devices; reclassification of single-use female condom, to be renamed single-use internal condom. Federal Register. September 27, 2018. Available at: https://www.federalregister.gov/documents/2018/09/27/2018-21044/obstetrical-and

-gynecological-devices-reclassification-of-single-use-female-condom-to-be
-renamed. Accessed May 22, 2019.

17. Rogers K. Why are condoms still a thing? *Vice.* February 13, 2015. Available at: https://www.vice.com/en_us/article/mgbx4x/why-are-condoms-still
-a-thing. Accessed June 1, 2019.

18. Bolan RK, Beymer MR, Weiss RE, Flynn RP, Leibowitz AA, Klausner JD. Doxycycline prophylaxis to reduce incident syphilis among HIV-infected men who have sex with men who continue to engage in high-risk sex: a randomized, controlled pilot study. *Sex Transm Dis.* 2015 Feb;42(2):98–103.

19. Molina JM, Charreau I, Chidiac C, Pialoux G, Cua E, Delaugerre C, Capitant C, Rojas-Castro D, Fonsart J, Bercot B, Bébéar C, Cotte L, Robineau O, Raffi F, Charbonneau P, Aslan A, Chas J, Niedbalski L, Spire B, Sagaon-Teyssier L, Carette D, Mestre SL, Doré V, Meyer L; ANRS IPERGAY Study Group. Postexposure prophylaxis with doxycycline to prevent sexually transmitted infections in men who have sex with men: an open-label randomised substudy of the ANRS IPERGAY trial. *Lancet Infect Dis.* 2018 Mar;18(3):308–17. doi: 10.1016/S1473-3099(17)30725-9. Epub 2017 Dec 8. PubMed PMID: 29229440.

20. Dubourg G, Raoult D. The challenges of preexposure prophylaxis for bacterial sexually transmitted infections. *Clin Microbiol Infect.* 2016 Sep;22(9):753–56. doi: 10.1016/j.cmi.2016.08.022. Epub 2016 Aug 29. Review. PubMed PMID: 27585939.

21. Fairley CK, Chow EPF. Doxycycline postexposure prophylaxis: let the debate begin. *Lancet Infect Dis.* 2018 Mar;18(3):233–4.

22. Spinelli MA, Scott HM, Vittinghoff E, Liu AY, Coleman K, Buchbinder SP. High interest in doxycycline for sexually transmitted infection postexposure prophylaxis in a multicity survey of men who have sex with men using a social networking application. *Sex Transm Dis.* 2019 Apr;46(4):e32–4.

23. Hook EW 3rd. 2008 Thomas Parran Award lecture. Translational research, STD control, and health disparities: a challenge and an opportunity. *Sex Transm Dis.* 2008 Dec;35(12):969–72.

24. Centers for Disease Control and Prevention. *A Public Health Approach for Advancing Sexual Health in the United States: Rationale and Options for Implementation, Meeting Report of an External Consultation.* Atlanta, GA: Centers for Disease Control and Prevention; 2010.

Index

About the Author

Ina Park, M.D., M.S., is an associate professor at the University of California, San Francisco, School of Medicine; a medical consultant at the Centers for Disease Control and Prevention, Division of STD Prevention; and medical director of the California Prevention Training Center. She holds degrees from the University of California, Berkeley, the David Geffen School of Medicine at UCLA, and the University of Minnesota School of Public Health. Recently, Park served as a coauthor of the 2020 CDC STD Treatment Guidelines and a contributor to the Department of Health and Human Services STI Federal Action Plan. A fierce advocate for public health, she lives in Berkeley, California, with her husband and two sons.